RANDOM
POLYMER
M●DELS

RANDOM
POLYMER
M●DELS

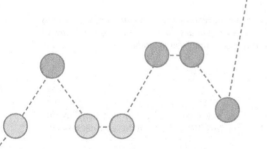

Giambattista Giacomin

Université Paris 7 – D. Diderot and
Laboratoire de Probabilités et Modèles Aléatoires,
CNRS – UMR7599, France

Imperial College Press

ICP

Published by

Imperial College Press
57 Shelton Street
Covent Garden
London WC2H 9HE

Distributed by

World Scientific Publishing Co. Pte. Ltd.
5 Toh Tuck Link, Singapore 596224
USA office: 27 Warren Street, Suite 401-402, Hackensack, NJ 07601
UK office: 57 Shelton Street, Covent Garden, London WC2H 9HE

British Library Cataloguing-in-Publication Data
A catalogue record for this book is available from the British Library.

RANDOM POLYMER MODELS

ISBN-13 978-1-86094-786-5
ISBN-10 1-86094-786-7

Printed in Singapore.

A Raika, a Lydia, e a Micol

nell'ordine in cui sono arrivate
e, per ora, in ordine di altezza

*Heavily weighs on me at times the burdensome reflection that I cannot
honestly say I am confident as to the exact shape of the once-seen,
oft-regretted Cube; and in my nightly visions the mysterious precept,
"Upward, not Northward," haunts me like a soul-devouring Sphinx. It is
part of the martyrdom which I endure for the cause of Truth that there
are seasons of mental weakness, when Cubes and Spheres flit away into
the background of scarce-possible existences; when the Land of Three
Dimensions seems almost as visionary as the Land of One or None; nay,
when even this hard wall that bars me from my freedom, these very
tablets on which I am writing, and all the substantial realities of Flatland
itself, appear no better than the offspring of a diseased imagination, or
the baseless fabric of a dream.*

From *Flatland: A Romance of Many Dimensions* by Edwin A. Abbott

Preface

This work deals with a well-defined class of probabilistic models. The focus is on polymers. More precisely I should say that the focus is on the equilibrium statistical mechanics of a class of polymers: dynamical and non-equilibrium phenomena are not treated, reflecting the fact that these directions are at the moment under-developed, at least from the viewpoint of the so called *rigorous results*. Moreover, only a subset of the world of polymer models is taken into consideration.

If I try to characterize such a subset, keeping in mind the motivations (that come from physics, biology, chemistry, material science, etc.), I end up with a list: $(1+d)$-dimensional pinning models, $(1+1)$-dimensional wetting models, adsorption models, copolymer models, DNA denaturation models, etc. and to each model in this list one should probably add the adjective *disordered*, since this work is mostly focused on disordered systems, even if non-disordered systems do play a central role. But such a list may at first appear quite *disordered* in itself...

If instead I take a purely mathematical standpoint, there is a totally natural thread connecting the models in the list: take a (persistent or terminating) renewal process and modify its law by giving rewards or penalties at the renewal epochs, possibly depending on time lapsed since the previous renewal epoch, and do that by introducing exponential, or *Boltzmann*, weights. On page 685 of his celebrated review paper *Walks, Walls, Wetting, and Melting* [Fisher (1984)] Michael E. Fisher writes:

"In fact, there is a rather simple but general mathematical mechanism which underlies a broad class of exactly soluble one-dimensional models which display phase transitions. This mechanism does not seem to be as well appreciated as it merits and it operates in a number of applications we wish to discuss."

The general mathematical mechanism exposed by Fisher is really the thread I mention previously, with the difference that in this book disorder is introduced and the soluble character of the models disappear. One of my aims is to look at this general mechanism from the perspective of Renewal Theory, making it (possibly) more appealing to mathematicians, linking it with a number of extremely sharp results developed in this beautiful branch of probability and setting up a framework in which disorder appears to be somewhat more friendly.

The book is organized as follows:

- Chapter 1 is a collection of (motivating, I hope) models, with partial solutions, that lead to the general framework given in Section 1.8. This general framework is definitely useful for proving some general results, but I found it more useful (and readable!) working out many of the results on more particular cases, even in the cases in which they hold in full generality.
- Chapter 2 and Chapter 3 deal with non-disordered models and the Renewal Theory approach is developed in detail: some of the arguments are relatively heavy. However the basic ideas of this approach are already in Chapter 1 and one can read Chapters 4 to 9, the chapters dealing with disordered models, almost skipping these two chapters. This may sound paradoxical, since in Chapters 5 to 8 the reader will find very many links (above all) to Chapter 2. But these parts of Chapter 2 can be read when the moment comes. I stress that I am only signaling this option and I am not advising against reading the book more sequentially.
- Chapter 4 deals with the problem of the existence of the free energy for disordered models. An elementary proof is worked out in detail and some *other* proofs are sketched. The reason for emphasizing *other* is simply that I find it troublesome talking about different proofs when they are all based on super-additivity.
- Chapter 5 and Chapter 6 focus, respectively, on the phase diagram of the pinning model and of the copolymer models, which are in a sense the two basic models (by combining them one finds the general framework). The two chapters can be read independently, but some of the arguments given for pinning are not repeated for copolymers (but of course their validity is stated in the bibliographic complements at the end of the chapter, with a sketch of how to modify the arguments).

- Chapter 7 and Chapter 8 deal with the properties of polymer paths. These are sharply different in the localized regime (Ch. 7) and in the delocalized one (Ch. 8). The two chapters are independent and they do not require a detailed knowledge of Ch. 5 and Ch. 6. In fact I believe that they can even be read directly after Ch. 4, but of course it may be somewhat reassuring knowing that a localized, or delocalized, regime does exist.

- Chapter 9 deals with computational approaches to analyze these systems. The focus is on both methods (algorithms, data analysis tools) and results.

- There are then three appendices. Appendix A is rather long and aims at making this work as self-contained as possible. There one will find a number of standard and not so standard results, mostly from probability theory (but not only). Appendix B collects some important technical estimates, and Appendix C is a very quick reminder that directed random walks in $(1 + 1)$-dimension are also *effective* models of interfaces and that some of our polymer models can in fact be reinterpreted in a different light.

In the sections in which the results are fully worked out (and these sections constitute about ninety percent of the book), the reader will find very few references, in fact almost none: the references are collected in the last section of each chapter, when discussing who did what and when, more things that have been done and what has not been done yet. I truly apologize for the (inevitable) omissions.

I have tried to make this book as self-contained as possible and the different chapters independent. My impression is that the diagram of dependences is something like the one in the figure, but of course I may be too optimistic.

And now a bit of history: I first met polymer models (precisely: copolymers) during my first post-doc (at the University of Zürich), when I met Erwin Bolthausen who suggested to me a number of very interesting problems that (I quote) *cannot be so difficult* (the responsibility for such a statement is entirely on Erwin's shoulders: some of these problems are the open problems in this book). At that time I drifted toward *not so difficult* problems not involving directly polymers, but that was not a choice: I have simply been driven by what I could solve first. However I kept playing with polymer questions and about three years ago I decided to make a serious effort, taking as a pretext also the fact that I had been invited by Franco

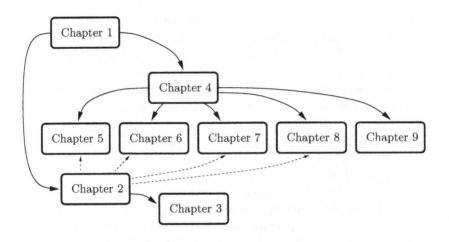

Diagram of possible reading orders of chapters. Dashed lines denote weaker links.

Flandoli to give a series of lectures at the Centro E. de Giorgi in Pisa during a trimester on *Interacting Particles and Computational Biology*. Polymer models and the *old questions* were ideal. On that occasion I started writing some notes, which then evolved through the years and through various graduate level mini-courses and courses I have given since. I have given full length math graduate courses at the University of Milano-Bicocca (I thank Daniela Bertacchi and Arrigo Cellina for the invitation) and at the Universities of Paris 6 and 7, where I work. I have also given a mini-course at the University of Pisa (I am grateful to Maurizio Pratelli and, again, to Franco Flandoli for inviting me) and right now, at the moment in which I am writing, at the Conference *Stochastic Processes in Mathematical Physics* that is held at Villa La Pietra (NYU, but in Florence). The latter event was an occasion to celebrate the sixtieth birthday of Michael Aizenman and Chuck Newman and this brought an extraordinary audience: I deeply thank the organizers and the scientific committee for having given me such an opportunity.

The lecture notes have gradually evolved thanks to the interaction with so many people through the years. However this book has been rewritten from scratch, starting about six months ago. The lecture notes still exist [Giacomin (2004)] and while they have, of course, a largely non-empty intersection with this book and contain substantially less material, a portion of them is not in here. This is due to choice of argument, and also line of presentation: in particular this book is really based on renewal processes and not on random walks (and, in particular, not on the simple random walk). Apart from generalizing and shifting the approach a bit, another reason for writing the book from scratch was to rethink and rewrite every argument with a more mature spirit (more mature with respect to the notes and to what I have published up to now on the subject). This of course increases the chance of mistakes and, certainly, of typos. But I hope that the reader will appreciate that essentially all the arguments presented here are somewhat, and at times substantially, different from what one already finds in the literature.

I am greatly indebted to many people for their direct and indirect contributions. Let me first mention the two people that up to now have marked most my scientific perspectives: Joel Lebowitz, my PhD advisor and constant presence later on, and Erwin Bolthausen, to whom I went as a post-doc right after my thesis. Then I want to thank my very close collaborators over the last three years (as may be apparent from the list of references): Thierry Bodineau, Francesco Caravenna, Fabio Toninelli and Lorenzo Zambotti. They were also among the people who helped me the most in reviewing the preliminary versions of this book. But of course the list of people from which I greatly benefited is much longer, each of them for different reasons. In such a list certainly belong: Ken Alexander, Jean Bertoin, Francis Comets, Jean-Dominique Deuschel, Ron Doney, Thomas Garel, Massimiliano Gubinelli, Frank den Hollander, Dima Ioffe, Cécile Monthus, Enzo Orlandini, Nicolas Pétrélis, Jacques Portes, Julien Sohier, Herbert Spohn, Yvan Velenik and Ofer Zeitouni. Six members of my lab (LPMA) are already mentioned above: let me however thank all the people of LPMA and also all the colleagues and friends at Paris 7 and Paris 6 for the very nice atmosphere that I have always felt in Jussieu-Chevaleret.

While working on polymers I rediscovered my passion for computer programming that characterized my pre-university life (later, theoretical stuff took over). I also discovered the fantastic world of General Public License software (I now truly believe that research and education cannot go on without free software), so it is a pleasure for me to take this opportunity to thank all the people who contribute to this world scale enterprise.

And to the three persons closest to me while writing, and closest to me *tout court*, I dedicate the book.

Giambattista Giacomin
Paris, June 30th 2006

Contents

Chapter 1

Random Polymer Models and their Applications

1.1 Random Polymers and (De)Localization Phenomena

Polymers are chemical compounds consisting *essentially* of repeating units, called *monomers* or, more properly, of *monomer units*. We will use indifferently *monomer* and *monomer unit*, for the precise meanings in polymer science see [Jenkins *et al.* (1996)]. The adverb *essentially* refers to the fact that in several instances the monomer units are not identical: as a matter of fact this situation is rather typical, but in many cases it is not inappropriate to neglect this aspect and to model the polymer as a chain of exactly repetitive units.

The polymers we are interested in are precisely the ones in which the units are not identical and this aspect cannot be neglected: at times these types of polymers are called *heteropolymers*. Two basic examples are sketched in Figure 1.1: the case of a biopolymer interacting with the membrane of a cell and the case of a *copolymer*, see the caption of Figure 1.1, close to the interface between two selective solvents. As they have been presented, only the second example deals with a heteropolymer. However, as it will be clear later, in our framework we will be free to re-interpret also the first case as a heteropolymer model.

Besides this basic aspect of the heterogeneous character of the chain, real polymers are complex objects on their own, typically fluctuating in a solvent and interacting with the molecules of the solvent as well as with other portion *of themselves*, the so called *self-interaction* or *excluded volume interaction* (two classical references on polymers and polymer models are [Flory (1953)] and [de Gennes (1979)]). However, in simplified models, polymers are often reduced to random walk paths on a lattice and the self-interaction is modeled by the *self-avoiding* constraint: a random walk path, self-

1

avoiding or not, is a snapshot of a polymer fluctuating in time. Self-avoiding random walks are extremely challenging models [Madras and Slade (1993)] and the same is true for other self-interacting walks [Bolthausen (2002)]: from a mathematician's viewpoint they still present many arduous open questions. Much more treatable, we could even say trivial, polymer models satisfying the self-avoiding constraint are *directed walk models*. These are simply random walks in which one component is deterministic and it is strictly increasing: for example if $\{S_n\}_{n=0,1,2,\dots}$ is a two dimensional simple random walk , *i.e.* $S_0 = (0,0)$ and $\{S_n - S_{n-1}\}_{n=1,2,\dots}$ is an IID sequence of random vectors uniformly distributed on $\{(0,1),(0,-1),(1,0),(-1,0)\}$, then $\{n, S_n\}_{n=0,1,\dots}$ is a directed walk on $\mathbb{Z}^3$.

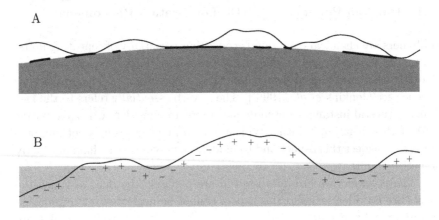

Fig. 1.1   A sketch of two systems that are representative of the problems we consider. In case A a polymer fluctuates in the proximity of an impenetrable (gray) region, for example the interior of a cell, to which is attracted in a non-homogeneous way, for example only the black sections of boundary (or membrane of the cell) are attractive for the polymer. In case B instead a polymer fluctuates in a medium constituted of two solvents (traditionally: oil and water). In this case the polymer is inhomogeneous: some monomer units are hydrophilic ($-$) and the other are hydrophobic ($+$). This type of polymers are often called *copolymers*. Energetically favored configurations tend to place $+$ monomers above the interface, and $-$ monomers below. In spite of this, the polymer fluctuates and perfect matchings are highly improbable for long chains, unless the temperature is very low. The same is true for case A: the *attractive* regions of the interface do not necessarily coincide with the contacts monomer-interface.

When, added to self-avoidance, the modeling of a precise physical system requires considering the interaction of a polymer with an environment or with other polymers (with few exceptions, we will not consider this second case, see however [Fisher (1984)]) it is often the case that this extra interaction increases the complexity to the point of making models based on self-avoiding random walks virtually untreatable. Building instead the model on a directed walk allows going further, or even much further, in the analysis. And it is possible that directed walk models still capture some fundamental features of real systems. This appears to be the case, see *e.g.* [Fisher (1984)], in several of the situations that we will consider in this work. Moreover, taking a mathematical standpoint (and not only...), directed models in interaction with external environments, above all if the interaction is inhomogeneous or *disordered*, often turn out to be extremely challenging (and yet not completely impenetrable). We cite for example the well known problem of a *directed polymer in random environment*, see *e.g.* [Comets *et al.* (2004)].

In the extremely wide world of polymers interacting with an environment we consider only a very specific subset. This choice in turn allows to go beyond directed walk models in a way that we now explain. For this, consider again Figure 1.1 and notice that in those two instances the interaction is either due to a contact polymer-interface or it depends on the location of the polymer on one side of the interface or the other. In reality in the first example there is also another interaction: the polymer cannot penetrate the membrane (we can see this as an infinite interaction that enters the game if the walk is below the membrane). In a way that we will understand much better later, these two models are two representatives of a large class of models that depend only on the sites of *contact* polymer-interface (the fact that the polymer in between two contacts is above or below the interface is a *degree of freedom* that can be integrated out). Therefore the directed walk turns out to be superfluous and the contact site process appears to be really more basic from a mathematical viewpoint and one can (greatly) generalize the original model by considering more general contact site processes (renewal processes: [Feller (1971)] and [Asmussen (2003)]). This may appear at first as an unmotivated generalization, but, as we will see, it is not the case.

But what is the phenomenology that we are after in these renewal based models? The keywords here are *localization* and (its opposite) *delocalization*. To explain this let us go back once again to Figure 1.1. Free polymers are long flexible chains fluctuating without interacting with the surrounding environment, but possibly subject to the excluded volume constraint. Associated to a free polymer of a given length there is a natural notion of *entropy*, that can be easily understood in lattice models since the entropy is simply the logarithm of the number of allowed polymer configurations (this is due to the fact that every allowed configuration is equiprobable). When they enter in contact with an environment the situation changes: the interaction is usually given in terms of an energy, that favors or penalizes certain configurations. In the extreme case in which the energy is extremely large, the interaction may simply lead to the suppression of certain configurations and the problem can be again described simply in term of entropy (for example when there is an impenetrable region). In general however one has really to play with an energy-entropy competition: in case A of Figure 1.1 the energy favors the *adhesion* of the polymer to the membrane, but it is not clear at all whether this effect prevails or not, since it has to overcome the entropic drive to exploring all possible polymer configurations. There is also the subtle effect coming from the presence of a forbidden region, like the interior of the cell in the example, that effectively leads to a repulsion of entropic origin. Note also that qualitatively the fact that the attracting *plaques* cover completely the membrane or that they cover it in a *scattered* way does not really affect the phenomenon in a substantial way. This makes all the difference with case B, the one of a *copolymer*, in which once again we have intuitively a localization effect due to the fact that there is an energetic gain in putting most of the hydrophobic monomers above the interface and most of the hydrophilic ones below: if all the monomers where hydrophobic, then there would be no reason to expect localization and, on the contrary, the energetic interaction would play in favor of the polymer living below the interface and thus effectively working in favor of delocalization.

This discussion has set forth the paradigm of our analysis: fluctuating linear chains interacting at a linear region. The result is either the localization of the polymer close to this region or delocalization, which roughly means that the polymer fluctuates freely except for taking care of avoiding precisely the region. This is actually not the complete picture: there is in fact also an intermediate situation, the critical one, appearing in particular when in our model there is one or more parameters that we can vary to pass from localization to delocalization. The critical regime is precisely the crossover point. It seems superfluous to stress that systems at criticality or near criticality are of particular interest for physicists and mathematicians. We will not be immune from such a *deformation*, but we will also pay attention to the more *boring* localized and delocalized regimes and we will see that they are not so boring, possibly because in several systems they are only partially, or even very partially, understood.

While we will cite and discuss several results taken from physics, biophysics and chemistry publications, all that follows is mathematics in the way it is presented and, of course, because the attention is toward rigorous results. The technical tools that we use most are the ones of probability theory, notably limit theorems and renewal theory. The reader may find in Appendix A reminders of a number of basic, and not always elementary, results that we use repeatedly.

Some recurrent notation and a (technical) suggestion

In what follows we use the symbol $\sim$ with two meanings:

(1) If f and g are two real valued functions we write $f(x) \overset{x \to x_0}{\sim} g(x)$, or simply $f(x) \sim g(x)$ when there is no danger of confusion, if $\lim_{x \to x_0} f(x)/g(x) = 1$.

(2) If X and Y are two random variables, $X \sim Y$ means that X and Y have the same law. We write also $X \sim \mathcal{N}(m, \sigma^2)$, for example, meaning that X is Gaussian with expectation m and variance σ^2.

The symbol $\asymp$ means instead asymptotic equivalence in the sense of Laplace. It is therefore reserved to positive functions, or to functions that are positive in a neighborhood of the point at which we are studying the behavior: $f(x) \overset{x \to x_0}{\asymp} g(x)$ if $\log(f(x)) \overset{x \to x_0}{\sim} \log(g(x))$.

The notation $a \wedge b$, respectively $a \vee b$, is sometimes used in order to spare room instead of $\min(a, b)$, respectively $\max(a, b)$, which is however preferred. For $a \in \mathbb{R} \setminus \mathbb{Z}$, $\lfloor a \rfloor$ (respectively $\lceil a \rceil$) is the largest integer smaller than a (respectively the smallest integer larger than a): if a is integer, $\lfloor a \rfloor = \lceil a \rceil = a$.

For $A \subset \mathbb{Z}$, θA is $A + 1$. When working on product spaces, for example $\mathbb{R}^{\mathbb{N}}$, we will use θ as translation operator: if $\zeta \in \mathbb{R}^{\mathbb{N}}$, then $\theta \zeta$ is an element of the same space and $(\theta \zeta)_n = \zeta_{n+1}$ for every $n \in \mathbb{N}$. With a habit which is not universally appreciated, we denote by $\mathbb{N}$ the set $\{1, 2, \ldots\}$.

The *technical suggestion* is instead about *slow variation*. Slowly varying functions, defined in Appendix A.4 (where one can find also a sum up of basic properties), appear in the definition (and in the analysis) of most of the models we consider as *corrections* to polynomial behaviors. These functions are often informally referred to as *logarithmic corrections* and in fact one of the most basic examples of slowly varying functions (at infinity) is $x \mapsto \log(1+x)$, $x \in (0, \infty)$. However if we replace $\log(1+x)$ with $(\log(1+x))^c$, any $c \in \mathbb{R}$, we still have a slowly varying functions and logarithmic functions are certainly not the only examples: $x \mapsto \exp(a(\log(1+x))^c)$

is slowly varying for every $a \in \mathbb{R}$ and every $c < 1$. But we suggest to keep in mind that the most elementary example of slowly varying function is a constant function or a function that tends (at infinity) to a positive constant (slowly varying functions are positive by definition and when we say positive we mean strictly positive): we call *trivial* a slowly varying function converging at infinity to a positive constant. So if the reader is not familiar with these functions, choosing them trivial in the definitions and in the arguments, with few exceptions, does not cause any trouble and allows to consider a class of models that is, anyway, reasonably wide.

1.2   A First Model: Pinning on a Defect Line

We start off by presenting a class of models capturing only very particular cases of the general framework outlined in the previous section. The polymer-environment interaction takes place only at the interface and it is homogeneous. These models are *completely solvable* and, in spite of the fact that they may look too simplistic, one gains substantial insight in working them out in detail. They will also come up repeatedly as *technical tools* in analyzing more complex models (*e.g.* for comparison arguments).

1.2.1   *Pinning a walk on a defect line*

In order to fix the ideas let us assume for the moment that $S := \{S_n\}_n$ is a symmetric random walk with increments taking values in $\{-1, 0, +1\}$. That is $S_0 = 0$ and $S_n = \sum_{j=1}^{n} X_j$, $n \in \mathbb{N}$, and $X := \{X_n\}_{n \in \mathbb{N}}$ is an IID sequence with $\mathbf{P}(X_1 = +1) = \mathbf{P}(X_1 = -1) = p/2 \in (0, 1/2)$ and $\mathbf{P}(X_1 = 0) = q > 0$, $p + q = 1$. The random variables X are defined on the probability space $(\Omega_S, \mathcal{F}, \mathbf{P})$. The law of S is $\mathbf{P}S^{-1}$, but we will simply say that the law of S is $\mathbf{P}$ when there is no risk of confusion. This process S will be from now on simply denoted as (p, q)-walk. The simple random walk is the $(1, 0)$-walk, that we exclude from our analysis (unless explicitly specified).

For any $\beta \in \mathbb{R}$ we consider the family of *Polymer measures* defined by

$$\frac{\mathrm{d}\mathbf{P}_{N,\beta}^{\mathrm{c}}}{\mathrm{d}\mathbf{P}}(S) = \frac{1}{Z_{N,\beta}^{\mathrm{c}}} \exp\left(\beta \sum_{n=1}^{N} \mathbf{1}_{S_n=0}\right) \mathbf{1}_{S_N=0}, \qquad (1.1)$$

for any $N \in \mathbb{N}$. The exponential term in the right-hand side is called *Boltzmann* factor and the quantity in the exponential is the *energy*. In statistical mechanics β is (temperature)$^{-1}$ and we can look at it as a parameter coupling the polymer to the environment. The *partition function*

$$Z^{\mathrm{c}}_{N,\beta} := \mathbf{E}\left[\exp\left(\beta\sum_{n=1}^{N} \mathbf{1}_{S_n=0}\right); S_N = 0\right],\qquad(1.2)$$

is the constant that makes $\mathbf{P}^{\mathrm{c}}_{N,\beta}$ a probability. Note the use of the semi-colon for restricting the expectation to a measurable subset. $\mathbf{P}^{\mathrm{c}}_{N,\beta}$ is the measure of a walk pinned at N, a *bridge*, rewarded (or penalized) β times the number of returns to zero up to and including the Nth step. We refer to Figure 1.2 for a visual image of the process and we stress that the index n of S is not interpreted as a time, but rather as a space parameter. More precisely, we are interested in the directed walk $\{(n, S_n)\}_n$, and essentially only for $n = 0, 1, \ldots, N$, even if it is often practical to consider all values of n.

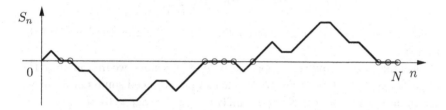

Fig. 1.2   A trajectory of the (p, q)-walk S or, rather, of the directed walk $\{(n, S_n)\}_n$ that we interpret as a (directed) polymer. The 10 contacts contributing to the energy are put in evidence ($N = 37$).

The superscript **c** stands for *constrained* and it refers to the presence of $\mathbf{1}_{S_N=0}$. This is a precise modeling choice, a boundary condition, and it is *a priori* as good as other boundary conditions, like for example *free* ones that is

$$\frac{\mathrm{d}\mathbf{P}^{\mathrm{f}}_{N,\beta}}{\mathrm{d}\mathbf{P}}(S) = \frac{1}{Z^{\mathrm{f}}_{N,\beta}}\exp\left(\beta\sum_{n=1}^{N}\mathbf{1}_{S_n=0}\right).\qquad(1.3)$$

In general, we use the following notation to restrict to $A \in \mathcal{F}$ a partition function:

$$Z^{\mathrm{f}}_{N,\beta}(A) := \mathbf{E}\left[\exp\left(\beta\sum_{n=1}^{N}\mathbf{1}_{S_n=0}\right); A\right],\qquad(1.4)$$

so that, for example, $Z^{\mathrm{c}}_{N,\beta} = Z^{\mathrm{f}}_{N,\beta}(S_N = 0)$.

In practice, the constrained case plays really an important technical role that will soon become clear. And exactly connected to this constraint we see why we do not allow working with the simple random walk ($q = 0$): we would have to require $N \in 2\mathbb{N}$. Of course this is not much of a problem, but we decide to privilege simpler notations. Possibly the reader is familiar with the elementary combinatorial formulas available for the simple random walk [Feller (1966), Chapter III] and those formulas are of course crucial for the intuition on random walks: but results that are, for all practical purposes, as precise are available for the (p, q)-walks (and beyond). A part of these results, the ones that we need, are recalled in Appendix A.6.

The importance of $Z_{N,\beta}^{c}$ (or $Z_{N,\beta}^{f}$), as a generating function, is quite clear. In particular we have

$$\frac{\partial}{\partial \beta} \frac{1}{N} \log Z_{N,\beta}^{c} = \mathbf{E}_{N,\beta}^{c} \left[\frac{1}{N} \sum_{n=1}^{N} \mathbf{1}_{S_n=0} \right], \tag{1.5}$$

which is the expected number of *contact* sites divided by the length of the polymer.

Aiming at studying the properties of long polymers, we first analyze the large N asymptotic behavior of $N^{-1} \log Z_{N,\beta}^{c}$. We need some notation: we set $\tau_0 := 0$ and, for $n \in \mathbb{N}$, $\tau_n := \inf\{m > \tau_{n-1} : S_m = 0\}$ if $\tau_{n-1} < \infty$, and $\tau_n := \infty$ otherwise. We set also $K(n) := \mathbf{P}(\tau_1 = n)$. It is well known, *cf.* Appendix A.6, that $K(n) \sim c_K n^{-3/2}$, as $n \to \infty$ and $c_K > 0$. Moreover $\sum_n K(n) = 1$, that is S is recurrent, and $\tau := \{\tau_i\}_{i=0,1,\ldots}$ is a sequence of $\mathbb{N}$–valued random variables (to be precise $\mathbb{N} \cup \{0\}$–valued, since $\tau_0 = 0$). The (strong) Markov property implies immediately that the sequence $\eta := \{\eta_i\}_{i \in \mathbb{N}}$, $\eta_i = \tau_i - \tau_{i-1}$, is IID. It will be useful to look at τ also as a random subset of $\mathbb{N} \cup \{0\}$, so in particular expressions like $N \in \tau$ and $\tau \cap A$ ($A \subset \mathbb{R}$) make sense.

Call $\mathrm{F}(\beta)$ the only solution of

$$\sum_n K(n) \exp(-\mathrm{F}(\beta)n) = \exp(-\beta), \tag{1.6}$$

when such a solution exists, that is for $\beta \geq 0$, and $\mathrm{F}(\beta) = 0$ when such a solution does not exist. Note that $\mathrm{F}(\beta) > 0$ for $\beta > 0$. It is immediate to see that $F : \mathbb{R} \to [0, \infty)$ is a non-decreasing continuous function.

Proposition 1.1 *For every β*

$$\mathrm{F}(\beta) = \lim_{N \to \infty} \frac{1}{N} \log Z_{N,\beta}^{c}. \tag{1.7}$$

We stress that for us a statement like (1.7) implicitly states also the existence of the limit.

Proof. It is just a matter of realizing that $Z_{N,\beta}^c$ is directly connected to the *mass renewal* function of a suitable renewal process (for generalities on renewal processes on $\mathbb{N}$ see Appendix A.5). More explicitly we write

$$Z_{N,\beta}^c = \sum_{n=1}^{N} \sum_{\substack{\ell \in \mathbb{N}^n: \\ \sum_{j=1}^{n} \ell_j = N}} \prod_{j=1}^{n} \exp(\beta) K(\ell_j). \tag{1.8}$$

Let us start with the case $\beta > 0$ and introduce the probability distribution $\widetilde{K}_\beta(n) = \exp(\beta) K(n) \exp(-\mathrm{F}(\beta)n)$. We may therefore write

$$Z_{N,\beta}^c = \exp\left(\mathrm{F}(\beta)N\right) \sum_{n=1}^{N} \sum_{\substack{\ell \in \mathbb{N}^n: \\ \sum_{j=1}^{n} \ell_j = N}} \prod_{j=1}^{n} \widetilde{K}_\beta(\ell_j) = \exp\left(\mathrm{F}(\beta)N\right) \widetilde{\mathbf{P}}_\beta(N \in \tau),$$

$$\tag{1.9}$$

where $\widetilde{\mathbf{P}}_\beta$ is the law of the (positive recurrent) renewal process with inter-arrival distribution $\widetilde{K}_\beta(\cdot)$. Of course $\widetilde{\mathbf{P}}_\beta(N \in \tau)$ is the mass renewal function of the process with inter-arrival distribution $\widetilde{K}_\beta(\cdot)$, computed in N, that is the probability that a walk with positive increments with law $\widetilde{K}_\beta(\cdot)$ passes by N. The Renewal Theorem (Theorem A.3) guarantees that $\widetilde{\mathbf{P}}_\beta(N \in \tau) \to 1/\sum_n n\widetilde{K}_\beta(n)$ for $N \to \infty$. Therefore

$$Z_{N,\beta}^c \sim \frac{1}{\sum_n n\widetilde{K}_\beta(n)} \exp\left(\mathrm{F}(\beta)N\right), \tag{1.10}$$

which of course implies $Z_{N,\beta}^c \asymp \exp(\mathrm{F}(\beta)N)$ in the same limit.

For $\beta \leq 0$, instead, $Z_{N,\beta}^c$ may be interpreted directly as a mass renewal function: $Z_{N,\beta}^c = \widetilde{\mathbf{P}}_\beta(N \in \tau)$. If $\beta = 0$, $\widetilde{\mathbf{P}}_\beta$ is a probability, but if $\beta < 0$ the discrete density $\widetilde{K}_\beta(\cdot) = \exp(\beta)K(\cdot)$ is a sub-probability, which of course may be regarded again as a probability, but on $\mathbb{N} \cup \{\infty\}$, by saying that the probability that an inter-arrival is equal to ∞ is $1 - \exp(\beta)$. In any case, the key feature is that for $\beta < 0$ the renewal process is terminating (or transient). The realization that

$$\exp(\beta)K(N) \leq Z_{N,\beta}^c \leq 1, \tag{1.11}$$

for $\beta \leq 0$ is immediate. This in particular says that $Z_{N,\beta}^c \overset{N \to \infty}{\asymp} 1$ for $\beta \leq 0$, and the proof is complete. $\qquad \square$

Remark 1.2 *If we set $E_j := \{S : \max\{n \leq N : S_n = 0\} = j\}$ for $j = 0, \ldots, N$, then $\cup_j E_j = \Omega_S$ and we may write, with the convention $Z^c_{0,\beta} := 1$, the formula*

$$Z^f_{N,\beta} = \sum_{j=0}^{N} Z^f_{N,\beta}(E_j) = \sum_{j=0}^{N} Z^c_{j,\beta} \sum_{n=N-j+1}^{\infty} K(n). \qquad (1.12)$$

By restricting the sum to $j = N$ we see that $Z^c_{N,\beta} \leq Z^f_{N,\beta}$. Moreover, by using the asymptotic properties of $K(\cdot)$, one finds a positive constant c such that $\sum_{n=N-j+1}^{\infty} K(n) \leq c(N-j)K(N-j) \leq cNK(N-j)$ for every j, so that

$$Z^c_{N,\beta} \leq Z^f_{N,\beta} \leq (1 + cN\exp(-\beta)) Z^c_{N,\beta}, \qquad (1.13)$$

since $Z^c_{N,\beta} = \sum_{j=0}^{N-1} Z^c_{j,\beta}K(N-j)\exp(\beta)$. Therefore we have also

$$\mathrm{F}(\beta) = \lim_{N\to\infty} \frac{1}{N}\log Z^f_{N,\beta}. \qquad (1.14)$$

Various properties of $\mathrm{F}(\cdot)$ follow now easily: $N^{-1}\log Z^c_{N,\cdot}$ is convex: this is proven for example by observing that its second derivative is a variance:

$$\frac{\partial^2}{\partial\beta^2}\frac{1}{N}\log Z^c_{N,\beta} = \frac{1}{N}\mathbf{E}^c_{N,\beta}\left[\left(\sum_{n=1}^{N}\mathbf{1}_{S_n=0} - \mathbf{E}^c_{N,\beta}\left(\sum_{n=1}^{N}\mathbf{1}_{S_n=0}\right)\right)^2\right],$$
$$(1.15)$$

Therefore $\mathrm{F}(\cdot)$ is convex and strictly increasing for $\beta > 0$. In this particular case one can compute exactly the Laplace transform of $K(\cdot)$, that is one can make the left-hand side of (1.6) explicit as a function of $\mathrm{F}(\beta)$, and invert the expression to find $\mathrm{F}(\beta)$. However, on a more abstract ground, the left-hand side of (1.6), for $\beta > 0$, and hence $\mathrm{F}(\beta) > 0$, is a real analytic function of $\mathrm{F}(\beta) > 0$ and therefore its inverse is real analytic too. So $\mathrm{F}(\cdot)$ is real analytic on the positive semi-axis. We conclude that we are dealing with a smooth function, except at 0, where of course it cannot be analytic. But which derivative (if any) is discontinuous?

In answering this question we can still avoid exact computations. In fact, let us first establish that for $b \searrow 0$

$$1 - \sum_n K(n)\exp(-bn) \sim 2c_K\sqrt{\pi b}. \qquad (1.16)$$

In fact, by summation by parts, the left-hand side is equal to

$$1 - \sum_{n=1}^{\infty} K(n) \exp(-bn) = (1 - \exp(-b)) \sum_{n=0}^{\infty} \left(\sum_{j=n+1}^{\infty} K(j) \right) \exp(-bn),$$
(1.17)

and, since $K(n) \sim c_K n^{-3/2}$, $\sum_{j=n+1}^{\infty} K(j) \sim 2c_K n^{-1/2}$. Note that of course we can neglect an arbitrarily large fixed number of terms in the sum in the right-hand side, say the first n_0 terms, and the error is $O(b)$. At the same time, for any $\varepsilon > 0$, if n_0 is sufficiently large, $c_K - \varepsilon \leq n^{3/2} K(n) \leq c_K + \varepsilon$ for $n \geq n_0$. By Riemann sums approximation we have that

$$\lim_{b \searrow 0} b^{1/2} \sum_{n \geq n_0} n^{-1/2} \exp(-bn) = \int_0^{\infty} \frac{1}{x^{1/2}} \exp(-x) \, dx = \sqrt{\pi}, \quad (1.18)$$

and (1.16) is established.

It is now just a matter of going back to (1.6): the only solution to that equation satisfies as $\beta \searrow 0$ (and therefore $F(\beta) \searrow 0$)

$$2c_K \sqrt{\pi F(\beta)} \sim \beta \tag{1.19}$$

that is

$$F(\beta) \sim c\beta^2, \quad \text{with } c = \frac{1}{4c_K^2 \pi}. \tag{1.20}$$

Therefore $F \in C^1$ in 0 and the second derivative has a jump discontinuity. This fact has an interesting interpretation in the light of (1.5) and of the properties of convex functions (see Appendix A.1.1): by convexity and C^1-regularity of $F(\cdot)$ the limit of the right-hand side of (1.5) exists for every value of β

$$N(\beta) := \lim_{N \to \infty} \mathbf{E}_{N,\beta}^{c} \left[\frac{1}{N} \sum_{n=1}^{N} 1_{S_n=0} \right], \tag{1.21}$$

and it is equal to $F'(\beta)$. We call $N(\cdot)$ *contact density* of the system. What we have seen is therefore that $N(\cdot)$ is smooth, except in zero. It takes the value 0 in the negative semi-axis, and it is positive in the positive semi-axis. Since $N'(\beta) = F''(\beta)$ for $\beta \neq 0$, we see that the derivative of the contact density has a jump in zero.

We are therefore in front of a first example of phase transition: the free energy $F(\cdot)$ is not analytic at $\beta = 0 =: \beta_c$ and, in general, we say that the system exhibits a phase transition at β if the free energy is not analytic at

β. We say also that a transition (at β_c) is of kth, $k \in \mathbb{N}$, order if the free energy is C^{k-1} at β_c, but not C^k. Therefore in the case under analysis the transition at β_c is of second order.

Another interesting elementary consequence of (1.6) is that $\lim_{\beta \to \infty} \mathrm{F}(\beta) - \beta = \log K(1)$. The convexity and regularity properties of $\mathrm{F}(\cdot)$, with the fact that $\mathrm{F}(\beta) \overset{\beta \to \infty}{\sim} \beta$, yields $\lim_{\beta \to \infty} \mathrm{F}'(\beta) = 1$. This result is rather intuitive: for β very large the polymer chain strongly binds to the defect line and the contact density is close to 1.

Figure 1.3 sums up what we have just obtained.

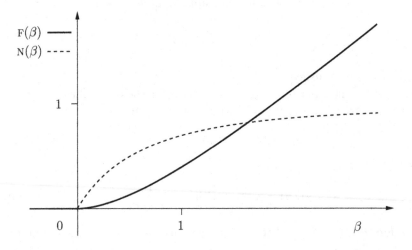

Fig. 1.3 Free energy and contact fraction for a homogeneous model of a directed polymer receiving rewards ($\beta > 0$) or penalties ($\beta < 0$) at a defect line. $\mathrm{F}(\cdot)$ is not analytical at $\beta = \beta_c := 0$. At β_c the free energy is C^1, but not C^2: there is a jump discontinuity in $\mathrm{F}''(\cdot) = \mathrm{N}'(\cdot)$ at β_c and the transition is therefore of second order.

Remark 1.3 *While in a sense the contact density viewpoint gives a very intuitive picture of the transition, we will systematically prefer to characterize the transition simply in terms of positivity of the free energy. The simple lower bound in (1.11), $\mathrm{F}(\beta) \geq 0$ allows to partition the parameter space, in this case the real line, into two subsets*

$$\mathcal{D} := \{\beta : \mathrm{F}(\beta) = 0\} \quad and \quad \mathcal{L} := \{\beta : \mathrm{F}(\beta) > 0\}, \tag{1.22}$$

where $\mathcal{D}$ stands for Delocalized and $\mathcal{L}$ for Localized. We call such a partition of the parameter space phase diagram. We stress that the possibility of *characterizing the phase diagram in terms of positivity of the free energy is peculiar to the particular models we consider. Moreover while in the simple completely solvable model we are considering we have proven that the free energy is not analytic only at $\beta_c = 0$ and therefore we are sure that the definition in (1.22) does identify the full phase diagram of the model (in principle the phase diagram should carry the information on all the singularities of the free energy), in general we will not be able to do so and therefore, at least in principle, such a decomposition may catch only a part of the details of the full phase diagram of the system.*

1.2.2 *Pinning more general processes on a defect line*

The random walk S defines the pinning model of Section 1.2.1, but the arguments really used only the distribution $K(\cdot)$ of the return times to zero. In particular we could generalize the model to any homogeneous Markov chain S: to avoid trivialities we should assume for example that $\mathbf{P}(S_n = 0) > 0$ for some n, but in general there is no reason to assume that 0 is recurrent for the chain.

Suggestive examples that show the generality of such a model include the case of a general lattice random walk in $(1 + 1)$ dimension, Figure 1.4(A), and the case of a directed walk in $1 + d$ dimension, that is the process $\{(n, S_n)\}_{n=0,1,\ldots}$, with S, like before, the partial sums of an IID sequence X, but this time X_1 is a discrete random variable taking values in $\mathbb{Z}^d$, with $\mathbf{P}(X_1 = 0) > 0$. Also in these cases we define $K(\cdot)$ as the distribution of the returns to the origin: of course it is very well possible that $\sum_n K(n) < 1$, like for $d \geq 3$ or if the walk is asymmetric.

But one could go much beyond by defining the model just in terms of the sub-probability distribution $K(\cdot)$ on $\mathbb{N}$, extended to $\mathbb{N} \cup \{\infty\}$ as we did before. Given an IID sequence $\eta := \{\eta_n\}_{n \in \mathbb{N}}$ with distribution $K(\cdot)$, we consider η as inter-arrival times and we define τ as partial sums process associated to η (*i.e.* $\tau_0 := 0$ and $\tau_j := \sum_{n=1}^{j} \eta_n$): τ is a discrete renewal process (Appendix A.5). The probability space in which we represent η is still denoted by $(\Omega_S, \mathcal{F}, \mathbf{P})$. The pinning model is of course defined as

$$\frac{d\mathbf{P}_{N,\beta}^a}{d\mathbf{P}}(\tau) = \frac{1}{Z_{N,\beta}^a} \exp\left(\beta \mathcal{N}_N(\tau)\right) \mathbf{1}_{\Omega_S^a(N)}, \qquad (1.23)$$

where $\mathcal{N}_N(\tau) := |\tau \cap \{1, \ldots, N\}|$ and $\Omega_S^c(N) = \{N \in \tau\}$, while $\Omega_S^f(N) = \Omega_S$.

All the models we will consider essentially boil down to return time

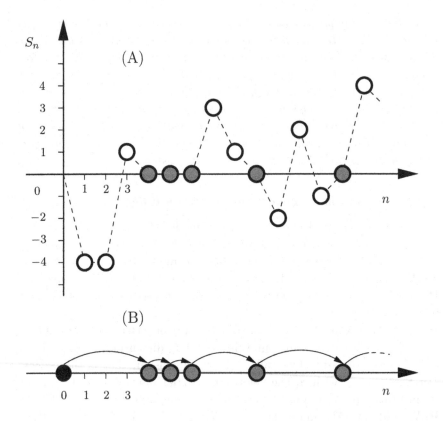

Fig. 1.4 A more general lattice walk: the jumps are not limited to taking values in $\{-1, 0, +1\}$. The walk can cross the x–axis, or defect line, without touching it (in gray the contacts with the defect line). The renewal points $\{\tau_0, \tau_1, \tau_2, \tau_3, \tau_4, \ldots\}$ are $\{0, 4, 5, 6, 9, \ldots\}$.

models, as it will be made clear in Section 1.8. And we will mostly focus on the class of $K(\cdot)$ distributions specified in the following definition.

Definition 1.4 Unless explicitly stated, the inter-arrival distribution, or return distribution, is a discrete sub-probability density $K(\cdot)$ on $\mathbb{N}$ which may be written as

$$K(n) = n^{-(1+\alpha)}L(n), \qquad (1.24)$$

with $\alpha \geq 0$ and $L : \mathbb{N} \to (0, \infty)$ a function that is slowly varying at infinity.

We stress that, with this definition, $K(n) > 0$ for every n. When useful, $K(\cdot)$ will be extended to a probability density on $\mathbb{N} \cup \{\infty\}$ by setting $K(\infty) = 1 - \Sigma_K$, $\Sigma_K := \sum_{n\in\mathbb{N}} K(n)$. In general we set $\overline{K}(n) :=$ $\sum_{j>n} K(j)$, where $n \in \mathbb{N} \cup \{0\}$ and in the sum the point ∞ is not included, so $\overline{K}(\infty) = 0$ and $\overline{K}(0) = \Sigma_K$. We also set $m_K := \sum_{n\in\mathbb{N}\cup\{\infty\}} nK(n) \in$ $(1,\infty]$ and of course $m_K = \sum_{n=0}^{\infty} \overline{K}(n)$ whenever $K(\infty) = 0$. We stress that in general $\sum_{j>n} \cdots$ means $\sum_{j\in\mathbb{N}:\, j>n} \cdots$ (∞ is excluded) and we will use expressions like $\sum_{n=1}^{\infty} a_n$ only if its interpretation is clear cut, that is when a_∞ is not defined, or when $a_\infty = 0$, or, still, when $\sum_{n\in\mathbb{N}} a_n = \infty$ and a_∞ is finite.

A summary on slow variation is given in Appendix A.4. For the time being we stress that the choice $K(\cdot) > 0$ has been made for technical convenience: the case of $K(\cdot)$ positive on a sub-lattice and/or only for n sufficiently large requires, at a conceptual level, only a minor additional effort. However notations and, at times, the arguments themselves become heavier, so we have made this choice for the sake of simplicity. Let us spell out one of the most important properties of $K(\cdot)$, which follows immediately from the definition of slowly varying function: for every choice of $\alpha_- < \alpha$ there exists $C_K > 0$ and for every $\alpha_+ > \alpha$ there exists $c_K > 0$ such that

$$\frac{c_K}{n^{1+\alpha_+}} \leq K(n) \leq \frac{C_K}{n^{1+\alpha_-}}, \tag{1.25}$$

for every $n \in \mathbb{N}$. Of course if $L(n)$ is trivial, that is if $L(n) \stackrel{n\to\infty}{\sim} c > 0$, we may choose $\alpha_- = \alpha_+ = \alpha$.

Remark 1.5 *Definition 1.4 excludes return time distributions that decay faster than power laws. We will come back to this issue after having presented various models, see Section 1.9.*

In Chapter 2 we will treat in detail the generalized pinning model we have just introduced. However most of the arguments in the proof of Proposition 1.1 are absolutely general and in any case $\mathrm{F}(\cdot)$ is nonnegative, it is convex, non-decreasing and it is determined by (1.6). Set $\beta_c = \inf\{\beta : \mathrm{F}(\beta) > 0\}$: from (1.6) one readily realizes that

$$\beta_c = -\log \Sigma_K \in [0,\infty). \tag{1.26}$$

This is a consequence of the fact that $\sum_n K(n) \exp(\varepsilon n) = +\infty$ for every $\varepsilon > 0$, so in particular in the cases of Definition 1.4 (but in reality (1.26) holds as soon as $\sum_n K(n) \exp(\varepsilon n) = +\infty$ for every $\varepsilon > 0$). By convexity and monotonicity $\mathrm{F}(\beta) > 0$ if and only if $\beta > \beta_c$, that is $\mathcal{L} = (\beta_c, \infty)$. In other words, $\beta_c = 0$ if and only if the renewal process τ, with law $\mathbf{P}$, is persistent (that is when S is recurrent, if S is present in the definition of the model: we will however use recurrent as synonymous with persistent also for τ). Moreover, by (1.6), the only singular point of $\mathrm{F}(\cdot)$ is β_c (see the argument before (1.16)). The rest that one can extract from (1.6) depends on finer details of $K(\cdot)$:

Proposition 1.6 *For the generalized pinning model, $\beta_c = -\log \Sigma_K$ and for $\beta \searrow \beta_c$*

$$\mathrm{F}(\beta) \sim (\beta - \beta_c)^{\max(1/\alpha,1)} \hat{L}(1/(\beta - \beta_c)), \tag{1.27}$$

with $\hat{L}(\cdot)$ a slowly varying function. $\hat{L}(\cdot)$ is trivial if $m_K < \infty$. Therefore the transition is of kth order if $\alpha \in (1/k, 1/(k-1))$. The order of the transition for $\alpha = 1/k$, $k \in \mathbb{N}$, is either k^{th} or $(k+1)^{th}$ and this depends on the slowly varying function $L(\cdot)$ that defines $K(\cdot)$.

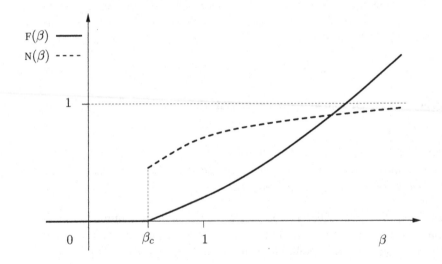

Fig. 1.5   Free energy and contact fraction for a homogeneous model with $\alpha > 1$ and $\Sigma_K < \infty$, so that $\beta_c = -\log \Sigma_K > 0$ and the contact fraction has a jump discontinuity at β_c, *i.e.* the transition is of first order.

Proposition 1.6 is a simplified version of Theorem 2.1. A complete proof is therefore postponed to Chapter 2, but observe that it is rather immediate to see that if $m_K < \infty$, then $\sum_n K(n)(1 - \exp(-bn)) \sim bm_K$ for $b \searrow 0$, so that (1.6) yields $\mathrm{F}(\beta)m_K \sim \Sigma_K(\beta - \beta_c)$, as $\beta \searrow \beta_c$. Therefore

$F(\beta) \sim c(\beta - \beta_c)$, with $c = \Sigma_K/m_K$. This means that $F(\cdot)$ is not C^1 at β_c and the transition is of first order.

We stress that we have used only $m_K < \infty$ and the explicit value of β_c, cf. (1.26), and these conditions are satisfied if $\alpha > 1$, cf. Definition 1.4, but possibly also when $\alpha = 1$ (that depends on $L(\cdot)$). And in reality the argument covers the case of general positive $K(\cdot)$ such that $\sum_n K(n) \exp(\varepsilon n) = +\infty$ for every $\varepsilon > 0$.

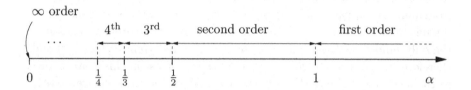

Fig. 1.6   The regularity of the transition. We stress that the order of the transition at the borderline points $1/j$, $j \in \mathbb{N}$, may be either j^{th} or $(j+1)^{\text{th}}$, in dependence of the slowly varying function $L(\cdot)$.

1.3   Entropic Repulsion and Wetting Phenomena

Important and often challenging questions arise in the analysis of interfaces between different phases or different materials, possibly connected to the presence of geometrical constraints like the presence of an impenetrable region. The statistical mechanics modeling of interfaces often resorts to the so called *reduced models*, that is to models in which the complexity is reduced by imposing to the trajectories of the process to be functions. In two dimensions, this naturally leads to random walk based models, see Appendix C.1.

Let us consider the following model

$$\frac{d\mathbf{P}_{N,\beta}^{c,+}}{d\mathbf{P}}(S) = \frac{1}{Z_{N,\beta}^{c,+}} \exp\left(\beta \sum_{n=1}^{N} \mathbf{1}_{S_n=0}\right) \mathbf{1}_{\{S_N=0\}\cap\Omega_S^+(N)}, \qquad (1.28)$$

with S a random walk with symmetric IID increments X and $\mathbf{P}(X_1 = j) \propto \exp(-V(j))$, for example $V(j) = |j|$, and $\Omega_S^+(N)$ is the subset of the trajectories S such that $S_n \geq 0$ for $n = 1, 2, \ldots, N$.

We start by rewriting $Z_{N,\beta}^{c,+}$ in strict analogy with (1.8):

$$Z_{N,\beta}^{c,+} = \sum_{n=1}^{N} \sum_{\substack{\ell \in \mathbb{N}^n: \\ \sum_{j=1}^{n} \ell_j = N}} \prod_{j=1}^{n} \exp(\beta) K(\ell_j), \qquad (1.29)$$

but this time $K(n) := \mathbf{P}\left(S_1 > 0, \ldots, S_{n-1} > 0, S_n = 0\right)$. $K(\cdot)$ is clearly a sub-probability and one can show, see (A.59) in Appendix A.6, that there exists a positive constant c_V^+ such that $K(n) \overset{n\to\infty}{\sim} c_V^+ n^{-3/2}$. Therefore, from a mathematical viewpoint, we are simply dealing with the generalized pinning model of Section 1.2.2: the transition is of second order and possibly the only *mild novelty* is that, in spite that S is recurrent, $K(\infty) > 0$ and therefore β_c, given by (1.26), is positive. The interest in this point is due to the fact that it shows that the repulsion effect due to the presence of a hard wall dominates, leading to delocalization, even in presence of a (small) strictly positive gain β per contact, in contrast with the situation without the wall constraint, where an arbitrarily small positive contact energy is sufficient for localization. This repulsion has an entirely entropic origin: only few of the trajectories in $\Omega_S^+(N)$ are close to the wall and they are therefore combinatorially (or entropically) penalized.

In order to understand a bit better this repulsion effect let us consider the model (1.29) with $\beta = 0$. In Section 2.2 we are going to show that $Z_{N,0}^{c,+} \overset{N\to\infty}{\sim} cN^{-3/2}$, for a suitable $c > 0$ (the precise value $3/2$ of the exponent is not important in what follows, but notice that it coincides with the decay exponent of $K(\cdot)$). From this it is then immediate to see that that for any $n \in \mathbb{N}$ and $\ell_1, \ldots, \ell_n \in \mathbb{N}$

$$\lim_{N\to\infty} \mathbf{P}_{N,0}^{c,+}\left(\tau_1 = \ell_1, \ldots, \tau_n = \ell_1 + \ldots + \ell_n\right) =$$

$$\prod_{j=1}^{n} K(\ell_j) \lim_{N\to\infty} \frac{Z_{N-\ell_1-\ldots-\ell_n,0}^{c,+}}{Z_{N,0}^{c,+}} = \prod_{j=1}^{n} K(\ell_j). \quad (1.30)$$

It is therefore clear that $\left\{\mathbf{P}_{N,0}^{c,+}\tau^{-1}\right\}_N$ converges weakly to the law of a (terminating) renewal process with inter-arrival distribution $K(\cdot)$. This means that the limit of $\left\{\mathbf{P}_{N,0}^{c,+}S^{-1}\right\}_N$ is transient and a walk in entropic repulsion touches the wall only a finite number of times before wandering off to infinity.

We will actually show that $Z_{N,\beta}^{c,+} \overset{N\to\infty}{\sim} cN^{-3/2}$ for any $\beta < \beta_c$ so the convergence in (1.30) holds also in this case and the limit process is still

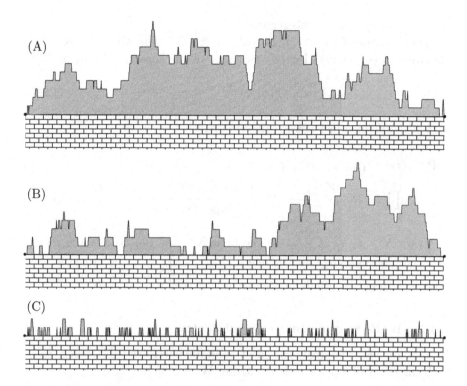

Fig. 1.7 The wetting model, based on a (p,q)-walk with $q = 0.7$. In (A) there is a typical delocalized trajectory ($\beta < \beta_c$): there are only a few returns, in fact $O(1)$ as $N \to \infty$, to the wall and all of them are close to the boundaries. In (C) instead there is a typical localized trajectory: the returns are frequent, in fact there is a density of pinned (or contact) points. In the wetting language, the random line is the interface between a liquid phase (below) and a gas phase (above). So in (A) is the system is in the *wet regime*, in the sense that the wall is covered with liquid, while in (C) it is partly dry, so we speak of *dry regime*. We will see in Chapter 2 that an intermediate scenario appears at $\beta = \beta_c$ (case (B)): there are $o(N)$ dry sites, but they may be found also in the bulk of the system.

a *strictly* delocalized walk (the return times have distribution $\exp(\beta)K(\cdot)$, which is not a probability, but only a sub-probability). A full analysis of the trajectories will be given in Chapter 2, but the simple considerations we have made are clearly pointing to a typical delocalized behavior as the one in Figure 1.7: only a few visits to the sites close to the walls and all these visits close to the endpoints.

Note that we are not considering the critical case $\beta = \beta_c$ that, by definition, is also delocalized, but in Figure 1.7(B) we give a quick preview of what is happening in this regime.

The situation for $\beta > \beta_c$ is really different. This time, as we have already seen (*cf.* (1.10)), by the classical Renewal Theorem (Theorem A.3)

$$Z_{N,\beta}^{\text{c},+} \exp(-\text{F}(\beta)N) \overset{N\to\infty}{\sim} 1/m_{\widetilde{K}_\beta} > 0, \tag{1.31}$$

with $\widetilde{K}_\beta(\cdot)$ defined just before (1.9). We have:

$$\mathbf{P}_{N,\beta}^{\text{c},+}(\tau_1 = \ell_1, \ldots, \tau_n = \ell_1 + \ldots + \ell_n)$$

$$= \prod_{j=1}^{n} \widetilde{K}_\beta(\ell_j) \frac{Z_{N-\ell_1-\ldots-\ell_n,\beta}^{\text{c},+} \exp(-\text{F}(\beta)(N - \ell_1 - \ldots - \ell_n))}{Z_{N,\beta}^{\text{c},+} \exp(-\text{F}(\beta)N)}, \tag{1.32}$$

for every $N > \ell_1 + \ldots + \ell_n$. In the limit $N \to \infty$ the ratio in the right-hand side converges to 1 and therefore $\left\{ \mathbf{P}_{N,\beta}^{\text{c},+} \tau^{-1} \right\}_N$ converges to a renewal process with exponentially integrable inter-arrival returns $\widetilde{K}_\beta(\cdot)$.

1.3.1   *Walls versus penetrable substrates*

It is interesting to compare the effect of an impenetrable substrate, a wall, with the effect of a penetrable substrate. With S as in (1.28)

$$\frac{\mathrm{d}\mathbf{P}_{N,\beta}^{\text{c},\lambda}}{\mathrm{d}\mathbf{P}}(S) = \frac{1}{Z_{N,\beta}^{\text{c},\lambda}} \exp\left(\beta \sum_{n=1}^{N} \mathbf{1}_{S_n=0} - \lambda \sum_{n=1}^{N} \mathbf{1}_{S_n<0} \right) \mathbf{1}_{\{S_N=0\}}, \tag{1.33}$$

for $\lambda \in [0,\infty]$. Note that the case $\lambda = \infty$ is the case of a hard, *i.e.* impenetrable, wall. The aim of this section is to argue that penetrable and impenetrable substrates give rise to the same qualitative behavior. Let us do that in the simple framework of (p,q)-walks. Again the key observation is that formula (1.8) holds also for $Z_{N,\beta}^{\text{c},\lambda}$: it is sufficient to replace $\exp(\beta)K(\ell_j)$ with $\exp(\beta)K(\ell_j)g_\lambda(\ell_j)$, where $g_\lambda(\ell) := (\exp(-\lambda(\ell-1)) + 1)/2$ and $K(\cdot)$ is the discrete probability density of the first return to the origin of a (p,q)-walk (note the connection with the model (1.28), where $K(\cdot)$ in fact coincides with $K(\cdot)g_\infty(\cdot)$).

It is therefore clear that the critical temperature β_c depends on λ, in fact $\beta_c = -\log \sum_\ell K(\ell)g_\lambda(\ell)$, and it is increasing in λ. But aside for this quantitative dependence, the arguments of the previous section go through with no change and, in particular, the delocalized trajectories behave in a way that differs from the trajectory 1.7(A) only in the fact that in the few short excursions close to the endpoints the polymer may visit the lower half-plane.

1.4 The Denaturation Transition: Poland–Scheraga Models

The DNA is a biopolymer that is often found in a double-stranded state: two *complementary* strands of DNA bind together in the well known helical form. *Complementary* refers to the fact that DNA is an inhomogeneous polymer, made up of monomer units of four different types (A, T, G, C, the four bases that carry the genetic code). Typically A binds with T and G binds with C so given a strand of DNA, the complementary strand is found by the substitutions A↔T and G↔C. Moreover A-T pairs carry two hydrogen bonds and they are therefore weaker than G-C pairs that carry three hydrogen bonds (mismatches are possible, but they are less stable). A first approximation is however simply to consider A-T and G-C bonds as equivalent and this leads to the model schematized in Figure 1.8: two homogeneous polymers interacting via an energy proportional to the number of contacts. Roughly, entropy works in favor of unbinding the two strands, while the energy gain tends to keep them together. The unbinding transition is called *denaturation transition* and, in real cases, it takes place at rather high temperatures (close to 90 degrees Celsius in *standard conditions* for an homogeneous G-C chain, and at about 65 degrees for an A-T chain: as one easily guesses, using homogeneous models for general chains may not be a good choice).

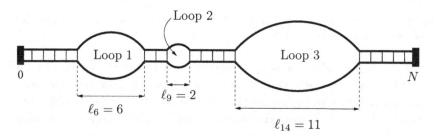

Fig. 1.8 A schematic view of the two strands of DNA. Some base pairs are in contact, and contribute an energy e, and some are detached, and they give no energy contribution. The base pairs that are not in contact form loops, 3 in the figure, and there is a contact at the end of each loop. In our definition of the Poland–Scheraga model $\ell_j = 1$ corresponds just to a contact (and effectively no loop). So the length of a loop is $2(\ell_j - 1)$ and a loop of length zero is a contact. In the example $N = 35$.

A first attempt to write down a more precise model may lead for example to consider two independent $(1 + 1)$–dimensional directed (p, q)-walks, see Figure 1.9, receiving a reward at the contact points. The walks may (for example) meet, but they cannot cross (we call this a *non-crossing constraint*). Of course the difference of two (p, q)-walks is still a random walk, precisely the walk S with

$$\mathbf{P}\left(S_1 = x\right) = \begin{cases} p^2/4 & \text{if } x = \pm 2, \\ pq & \text{if } x = \pm 1, \\ (p^2/2) + q^2 & \text{if } x = 0. \end{cases} \tag{1.34}$$

The non-crossing constraint therefore becomes a hard wall constraint for the walk S and we are effectively dealing with the wetting model of Section 1.3 (with a particular choice of $V(\cdot)$). In absence of constraint, the model falls of course in the class of models discussed in Section 1.2 and, as we have seen, the general models introduced in Section 1.2 include also the wetting models. In particular we have seen that the hard wall constraint induces simply a shift of the critical point.

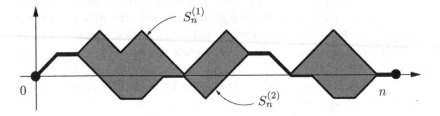

Fig. 1.9 Two (p, q)-walks, $S^{(1)}$ and $S^{(2)}$. They can touch but they cannot cross. Both walks are pinned at the endpoints.

One can generalize the simple random walk model by considering for example a more realistic $(1 + 2)$–dimensional model: note that the non-crossing constraint looses meaning in this case, but in any case the difference of two random walks is still a random walk and the model reduces to pinning on a defect line.

The substantial difference between the $(1 + 1)$ and the $(1 + 2)$–dimensional case is in the asymptotic behavior of $K(\cdot)$, governed respectively by $\alpha = 1/2$ and $\alpha = 1$ (*cf.* Appendix A.6) and we have seen how sensible to the values of α the critical behavior is.

Before going into this issue, which is at the heart of much of the work on DNA denaturation, let us introduce a general model that includes the cases discussed up to now: the Poland–Scheraga model. Let us do that by sticking first to the notation that is (more or less) standard in biophysics: we will then make the link with our notations on the way.

The Poland–Scheraga model is introduced by assigning the so called *loop entropy* $\mathcal{S}(\ell)$ which is taken with the property

$$\mathcal{S}(\ell) \overset{\ell \to \infty}{\sim} \sigma \frac{\mu^\ell}{\ell^c}, \tag{1.35}$$

where $\mu > 1$ is a geometric factor (reputed *non-universal* and usually considered of little importance), c is called *loop closure exponent* and it is typically chosen larger than $3/2$, $\sigma > 0$ is called *cooperativity parameter*. As a matter of fact the choice that is often made is $\mathcal{S}(\ell)$ equal to the right-hand side of (1.35) for $\ell \geq 2$ and $\mathcal{S}(1) = \mu$ (in a sense, $\sigma = 1$ if $\ell = 1$): since σ is often chosen rather small, for example $\sigma = 10^{-5}$, this particular choice of $\mathcal{S}(\cdot)$ has the effect of sharpening the unbinding transition (we will come back to this issue below). In any case we will assume that $\mathcal{S}(\ell) > 0$ for every $\ell \in \mathbb{N}$ and, without loss of generality, that $\sum_\ell \mathcal{S}(\ell)/\mu^\ell = 1$. We choose the boundary conditions by deciding that the first and the last monomer couple are in a bound state. Therefore the probability of observing a configuration $\underline{\ell} \in \mathbb{N}^n$, $n \leq N$, is equal to

$$\mathbf{P}_N^{\mathrm{PS}}(\underline{\ell}) := \frac{1}{Z_N^{\mathrm{PS}}} \exp\left(-\beta \mathrm{E} n\right) \prod_{j=1}^n \mathcal{S}\left(\ell_j\right) \tag{1.36}$$

if $\sum_{j=1}^n \ell_j = N$ and $\mathbf{P}_N^{\mathrm{PS}}(\underline{\ell}) = 0$ otherwise. In terms of loops, one should interpret the $\underline{\ell}$ configuration by thinking that $\ell_j = 1$ corresponds to a bound pair and $\ell_j > 1$ corresponds to a loop involving ℓ_j pairs. In (1.36), β is the reciprocal of the temperature and $\mathrm{E} < 0$ is the binding energy.

Quite a bit of insight is gained once we observe that

$$Z_N^{\mathrm{PS}} \mu^{-N} = \sum_{n=1}^N \sum_{\substack{\underline{\ell} \in \mathbb{N}^n: \\ \sum_{j=1}^n \ell_j = N}} \prod_{j=1}^n \exp(-\beta \mathrm{E}) K(\ell_j), \tag{1.37}$$

where $K(\ell) = \mathcal{S}(\ell)/\mu^\ell$. Notice that, by assumption, $K(\cdot)$ is a probability satisfying the conditions of Definition 1.4 (with $\alpha = c - 1$ and $\lim_{n \to \infty} L(n) = \sigma$). Notice also that if $\sum_\ell \mathcal{S}(\ell)/\mu^\ell \neq 1$, one can normalize the term simply by adding a constant to $\beta \mathrm{E}$. Therefore $Z_N^{\mathrm{PS}}/\mu^N = Z_{N,-\beta \mathrm{E}}^{\mathrm{c}}$

and the Poland–Scheraga model is equivalent to the homogeneous model of pinning on a defect line, in particular

$$\mathrm{F}^{\mathrm{PS}}(\beta) = \log \mu + \mathrm{F}(-\beta \mathrm{E}), \tag{1.38}$$

with $\mathrm{F}^{\mathrm{PS}}(\cdot)$ the free energy of the Poland–Scheraga model. The denaturation is of course characterized by $\mathrm{F}^{\mathrm{PS}}(\beta) = \log \mu$, while $\mathrm{F}^{\mathrm{PS}}(\beta) > \log \mu$ is the mark of the bound state.

Formula (1.38) and Proposition 1.6, visually represented in Figure 1.6, allow to draw some important conclusions.

- First of all that the transition is of first order only if $c > 2$: in particular if the loop entropy is the one taken from $(1 + d)$–dimensional directed models, $d = 1$ or 2, the transition is of higher order (for the case $d = 1$ we have seen that $c = 3/2$ and the transition is of second order, while $\alpha = 0$, so $c = 1$, for any aperiodic random walk with centered and finite variance increments in $d = 2$, see Appendix A.6, and therefore the transition is of infinite order, see Proposition 1.6 and Theorem 2.1). Considering two indirected walks in three dimensions (*a priori* a reasonable choice) does not help: the underlying renewal is given by the meeting times of two three-dimensional walks or, equivalently, by the returns to the origin of a three-dimensional walk, so that $c = 3/2$ as in one dimension (of course the renewal is terminating, but this simply leads to a shift of the critical temperature). In order to observe a first order transition one should go up to dimension 5 ($c = d/2$, see Appendix A.6), which does not seem very reasonable.

- There is a reason why we are insisting on looking for a first order transition. In fact what is observed from real DNA is something that may suggest that the transition is first order. In reality the situation is a bit more complex and what one really observes is what is called a *multistep* transition. It is hard to define precisely what this means, but, if we accept to keep at a rather intuitive level, it is quite clear that A-T rich regions unbind at lower temperature than G-C rich regions. Added to that, real experiments deal with N finite (and never too large, normally up to $\approx 10^6$ base pairs). So one should possibly accept the idea of giving up the infinite volume limit. However, there is evidence that unbinding of sub-regions of the double strand is quite a sharp phenomenon: this is relatively clear from observations on intermediate length strands (10^3 to 10^4 base pairs), while for longer (or shorter) chains the phenomenon is less evident.

- Higher values of c may arise if one adds self-avoiding constraints. For example one could add the self-avoiding constraint of the two strands inside a loop (of course one is using a lattice model and in this case there is a contact when the two strands are on nearest neighbor sites). This suggests an exponent $c \approx 1.75$, which is not enough for a first order transition. More recently it has been claimed that including the self-avoiding constraint of a loop with the rest of the chain leads to a value of c between 2.10 and 2.20: the transition is first order. In agreement with a certain amount of the literature we settle for $c = 2.15$. From a modeling viewpoint accepting on one hand the leading role of global self-avoiding constraints, keeping on the other the basic renewal structure, is rather troublesome and probably requires further work to clarify the precise extent of the approximations used up to now.

- The role of (small values of) the cooperativity parameter σ becomes now clear. Let us analyze this question by choosing:

$$K(\ell) \, = \, \mathcal{S}(\ell)\mu^{-\ell} \, = \, \begin{cases} \sigma\ell^{-c} & \text{if } \ell = 2, 3, \ldots, \\ 1 - \sigma\mathrm{C}_c & \text{if } \ell = 1, \end{cases} \tag{1.39}$$

where $\mathrm{C}_c = \sum_{n=2}^{\infty} n^{-c} \in (0, \infty)$. Small values of σ penalize the opening of loops. However with our definitions there is no effect on the critical β of the model, which is in any case zero. So the net effect is necessarily a sharpening of the transition: there is too little entropy gain in opening a loop, so before the denaturation transition the two strands are very tightly bound and, at the transition, it is more efficient to open *quickly* very large loops to act against the small value of σ. Observe moreover that even if we stick more closely to the definition of the model used in the application and if we redefine $K(1) \in (0, \infty)$ as an arbitrary fixed constant, then $\beta_c|\mathrm{E}| = \log(K(1) + \sigma\mathrm{C}_c)$ which behaves like $\log(K(1)) + \sigma\mathrm{C}_c/K(1)$ for small σ, so the effect of σ on β_c is mild.

Of course the next step is to introduce inhomogeneities in the model, for example in terms of two different energies $\mathrm{E_{AT}}$ and $\mathrm{E_{GC}}$. But this corresponds to considering an inhomogeneous pinning model (and we will come back to this at length).

1.5   Force Induced Unzipping

Delocalization may arise as a result of the action of an external force. This is motivated by several real experiences, we mention in particular some recent experiments on mechanically induced unbinding (or unzipping) of a double strand of DNA via experiments at a molecular level (see Section 1.10 for literature and further details). From the arguments in Section 1.4 it is clear that we can directly look at the reduced model of pulling a polymer out of an attractive defect line. The model is the following:

$$\frac{d\mathbf{P}^F_{N,\beta}}{d\mathbf{P}}(S) = \frac{1}{Z^F_{N,\beta}} \exp\left(\beta \sum_{n=1}^{N} \mathbf{1}_{S_n=0} + FS_N\right), \qquad (1.40)$$

where F is the force applied to the endpoint of the polymer. By symmetry, we may assume $F > 0$ without loss of generality.

We have the following result:

Theorem 1.7   *The free energy* $\mathrm{F}_{\mathrm{pull}}(\beta, F)$, *that is the limit of the sequence* $\left\{(1/N)\log Z^F_{N,\beta}\right\}_N$, *exists. Moreover we have the formula*

$$\mathrm{F}_{\mathrm{pull}}(\beta, F) = \max\left(\mathrm{F}(\beta), \mathrm{F}_{\mathrm{unzip}}(F)\right), \qquad (1.41)$$

where $\mathrm{F}_{\mathrm{unzip}}(F) := \log\left(p\cosh(F) + q\right)$.

Before proving this result, we observe that for fixed $\beta > 0$ it is natural to define $F_c(\beta) = \mathrm{F}^{-1}_{\mathrm{unzip}}(\mathrm{F}(\beta))$:

$$\mathrm{F}_{\mathrm{pull}}(\beta, F) = \begin{cases} \mathrm{F}(\beta) & \text{if } F \le F_c(\beta), \\ \mathrm{F}_{\mathrm{unzip}}(F) & \text{if } F \ge F_c(\beta). \end{cases} \qquad (1.42)$$

Since $\mathrm{F}(\cdot)$ is analytic outside of 0 and since $\mathrm{F}_{\mathrm{unzip}}(\cdot)$ is analytical on $(0, \infty)$, then $\mathrm{F}_{\mathrm{pull}} : (0, \infty)^2 \to \mathbb{R}$ is analytical on $(0, \infty)^2 \setminus \{(\beta, F) : F = F_c(\beta)\}$. Moreover the contact density, $\partial_\beta \mathrm{F}_{\mathrm{pull}}(\beta, F)$, is zero if $F > F_c(\beta)$ and it is positive if $F < F_c(\beta)$, so this is clearly a localization-delocalization transition. And this transition is of first order, since the contact density is independent of the value of F as long as it is smaller than $F_c(\beta)$.

By taking the derivative of $\mathrm{F}_{\mathrm{pull}}(\beta, \cdot)$ one readily sees that

$$\lim_{N \to \infty} \mathbf{E}^F_{N,\beta}[S_N/N] = \begin{cases} p\sinh(F)/(p\cosh(F) + q) & \text{if } F > F_c(\beta), \\ 0 & \text{if } F < F_c(\beta), \end{cases} \qquad (1.43)$$

and we see that also from this viewpoint the transition is discontinuous (*i.e.* first order).

As it will be clear from the proof Theorem 1.7 is rather general (and it goes even well beyond the homogeneous set-up, *cf.* Section 1.10): one can for example state it for general pinning models, but one has to add the information on the entropic cost due to the last (incomplete) excursion, an information that is clearly not contained in the renewal process.

Lemma 1.8 *If $\{a_n\}_n$ and $\{b_n\}_n$ are two sequences of positive numbers (except for finitely many of them that can take value zero) such that $a_n \asymp \exp(an)$ and $b_n \asymp \exp(bn)$ for $n \to \infty$, with a and b two real numbers, then*

$$\sum_{n=0}^{N} a_n b_{N-n} \overset{N \to \infty}{\asymp} \exp(\max(a,b)N). \tag{1.44}$$

Proof. Assume without loss of generality that $a \geq b$ and remark that the left-hand side in (1.44) is equal to $\exp(aN)$ times $\sum_{n=0}^{N} \exp(-an)a_n \exp(-a(N-n))b_{N-n}$. The last expression is the discrete convolution of two sequences with sub-exponential growth. This proves that $\lim_N (1/N) \log \sum_{n=0}^{N} a_n b_{N-n} \leq a$. For the lower bound just observe that the left-hand side in (1.44) is larger than $b_k a_{N-k}$, with $k = \min\{n : b_n > 0\}$. $\qquad\square$

Proof of Theorem 1.7. The proof is based on the elementary formula

$$Z_{N,\beta}^{F} = \sum_{n=0}^{N} Z_{n,\beta}^{c} Z_{N-n}^{+}(F), \tag{1.45}$$

with

$$Z_n^{+}(F) := \sum_{k \in \mathbb{Z}} \exp(Fk) \mathbf{P}\left(S_j > 0,\ j = 1, 2, \dots, n-1,\ S_n = |k|\right), \tag{1.46}$$

and $Z_0^{+}(F) = 1$. By Proposition 1.1 and Lemma 1.8, the proof is reduced to showing that

$$\lim_{n \to \infty} \frac{1}{n} \log Z_n^{+}(F) = \mathrm{F}_{\mathrm{unzip}}(F). \tag{1.47}$$

This is a computation. For $k > 0$ we have

$$\mathbf{P}(S_j > 0, \, j = 1, 2, \ldots, n-1, \, S_n = k) =$$
$$\frac{p}{2}\left[\mathbf{P}(S_{n-1} = k-1) - \mathbf{P}(S_{n-1} = k+1)\right]. \quad (1.48)$$

This formula follows from the reflection principle (the proof is detailed in Appendix A.6). It is useful to remark that the sum in (1.46) may be restricted to positive k's (this leads to an error term $O(1)$). Therefore

$$
\begin{aligned}
Z_n^+(F) &= \frac{p}{2}\sum_{k \in \mathbb{N}} \exp(Fk)\left[\mathbf{P}(S_{n-1} = k-1) - \mathbf{P}(S_{n-1} = k+1)\right] + O(1) \\
&= p\sinh(F)\sum_{k \in \mathbb{N}} \exp(Fk)\mathbf{P}(S_{n-1} = k) + O(1) \\
&= p\sinh(F)\sum_{k \in \mathbb{Z}} \exp(Fk)\mathbf{P}(S_{n-1} = k) + O(1) \\
&= p\sinh(F)\mathbf{E}\left[\exp(FS_{n-1})\right] + O(1) \\
&= p\sinh(F)\left(p\cosh(F) + q\right)^{n-1} + O(1),
\end{aligned}
$$
$$(1.49)$$

where the $O(1)$ terms change from line to line. This establishes (1.47) and the proof is complete. $\qquad\qquad\qquad\qquad\qquad\qquad\qquad\qquad\qquad\quad\square$

So Theorem 1.7 tells us that a force applied to the endpoint of the polymeric chain has either a very drastic effect or (almost) no effect at all: either the polymer essentially behaves ballistically, as suggested by (1.43), or the force has a negligible effect. It is probably important to remark at this point that the effect of pinning the endpoint at a height which is proportional to N, that is considering the model with partition function $Z_{N,\beta}^f(S_N = \lfloor aN \rfloor)$ for $a \in (0, 1]$, leads to a rather different behavior. This can be analyzed again by conditioning to the last hitting point on the defect line, as in (1.45), and this time $Z_n^+(F)$ has to be replaced by a pure entropic term $p_{\lfloor aN \rfloor}(n)$ given by the probability that a (p, q)-walk stays positive up for n steps and that it is at height $\lfloor aN \rfloor$ after precisely n steps. By the reflection principle (Proposition A.9) we have

$$p_M(n) = \frac{p}{2}\left[\mathbf{P}(S_{n-1} = M-1) - \mathbf{P}(S_{n-1} = M+1)\right]. \quad (1.50)$$

We will not give the details of the computation here, but the net result is that the free energy of this model is

$$\max_{\gamma \in [0, 1-a]} \left[\gamma \mathrm{F}(\beta) - (1-\gamma) \Sigma_{S_1}(a/(1-\gamma)) \right], \qquad (1.51)$$

where $\Sigma_{S_1}(\cdot)$ is the (Cramer) Large Deviations functional of the increments of the (p, q)-walk, which of course diverges (in a discontinuous fashion) at 1 (that explains the restriction of the maximum to $\gamma \in [0, 1-a]$). A quick look at this problem shows that the maximizer $\gamma_* := \gamma_*(a, \beta)$ belongs to $(0, 1-a]$. The behavior of this system is actually sketched in Figure 1.10(B).

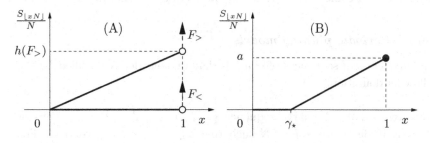

Fig. 1.10   In (A) a polymer with a pinning potential at a defect line to which a force has been applied. The function $F \mapsto h(F)$ is defined in (1.43). We have chosen $F_> > F_c(\beta)$ and $F_< < F_c(\beta)$, so $h(F_<) = 0$. In (B) instead the polymer endpoint is pinned at height $\lfloor aN \rfloor$. In both cases what we plot is what one sees at large N, after rescaling by N. In particular fluctuations are suppressed.

1.6   Inhomogeneous Charge Distributions: Copolymers and Pinning

We introduce inhomogeneities in the polymer chain, or at the interface or defect line, by assigning to each monomer, or to each interface site, a *charge*, that is a real number that will determine quantitatively the interaction of the chain with the surrounding environment. For example, an inhomogeneous pinning model is the following:

$$\frac{\mathrm{d}\mathbf{P}_{N,\omega,\beta}^a}{\mathrm{d}\mathbf{P}}(\tau) = \frac{1}{Z_{N,\omega,\beta}^a} \exp\left(\beta \sum_{n \in \tau, n \le N} \omega_n \right) \mathbf{1}_{\Omega_S^a(N)}, \qquad (1.52)$$

with $\omega = \{\omega_n\}_{n \in \mathbb{N}}$ a sequence of real numbers and for $a = \mathtt{c}$ or $\mathtt{f}$, as usual. For inhomogeneous models it is at times practical to use, for $M \in \mathbb{N} \cup \{0\}$

and $M < N$ the notation

$$Z^a_{M,N,\omega,\beta} := Z^a_{N-M,\theta^M\omega,\beta},\qquad(1.53)$$

and $Z^a_{M,N,\omega,\beta} = 1$ if $M = N$.

With a binary choice of the values of ω_n, that is $\omega_n \in \{a, b\}$, $a < b$, corresponding to the interaction energies of A–T and C–G couples, in view of the discussion in Section 1.4, $\mathbf{P}^a_{N,\omega,\beta}$ is an inhomogeneous Poland–Scheraga model, which gives a more realistic model for the DNA denaturation transition (we will come back to this issue in Section 1.10 and in Section 5.7).

1.6.1 *Periodic pinning models*

The first class of sequences of charges that we consider is specified by the following definition.

Definition 1.9 A charge sequence $\{\omega_n\}_n$ is weakly inhomogeneous or periodic if there exists $T \in \mathbb{N}$ such that $\omega_{n+T} = \omega_n$ for every n. The minimal such T is called the period and it is denoted by $T(\omega)$. A periodic sequence is said centered if $\sum^T_{n=1} \omega_n = 0$.

Localization and delocalization for a periodic pinning model may be characterized once again by looking at the free energy. As we will explain in Chapter 3, it is possible to generalize the renewal theory approach introduced in Section 1.2, however the algebra is substantially more complex and the point process hidden behind periodic models is not a standard renewal, but rather a *Markov renewal* (see Chapter 3). This will allow precise computations, but for the moment we observe that:

Proposition 1.10 *If ω is a periodic sequence of charges then the limit of the sequence $\left\{N^{-1}\log Z^c_{N,\omega,\beta}\right\}_N$ exists and we denote it by $\mathrm{F}_\omega(\beta)$. $\mathrm{F}_\omega(\cdot)$ takes values in $[0,\infty)$ and it is convex. Moreover $\mathrm{F}_{\theta\omega}(\cdot) = \mathrm{F}_\omega(\cdot)$ and $\mathrm{F}_\omega(\cdot) = \mathrm{F}_{\widetilde\omega}(\cdot)$, with $(\widetilde\omega)_n = \omega_{T(\omega)-n}$ for $n = 0, 1, \dots T(\omega) - 1$ and $\widetilde\omega$ is extended by periodicity.*

Proof. First of all set $T = T(\omega)$ (we assume $T \geq 2$, otherwise the model is homogeneous) and observe that for every $n \in \mathbb{N}$ and $m \in \mathbb{N} \cup \{0\}$

$$Z^c_{nT,\omega,\beta} \geq Z^c_{nT,\omega,\beta}(mT \in \tau) = Z^c_{mT,\omega,\beta}Z^c_{(n-m)T,\theta^{mT}\omega,\beta},\qquad(1.54)$$

and as usual $Z^{\mathrm{c}}_{0,\omega,\beta} = 1$. Since $\theta^{mT}\omega = \omega$, one immediately sees that the sequence $\left\{\log Z^{\mathrm{c}}_{nT,\omega,\beta}\right\}_n$ is super-additive so that, *cf.* Theorem A.12, the limit $\mathrm{F}_\omega(\beta)$ of $\left\{(nT)^{-1}\log Z^{\mathrm{c}}_{nT,\omega,\beta}\right\}_n$ exists. The limit is clearly bounded above by $\beta \max_n(\max(\omega_n, 0))$. On the other hand by restricting the computation of the partition function to the trajectories τ such that $\tau \cap \{1, \ldots N-1\} = \emptyset$ we see that $Z^{\mathrm{c}}_{N,\omega,\beta} \geq \exp(\beta\omega_N)K(N) \overset{N\to\infty}{\asymp} 1$, and so $\mathrm{F}_\omega(\beta) \geq 0$. The convexity of $\mathrm{F}_\omega(\cdot)$ follows from the convexity of $\log Z^{\mathrm{c}}_{nT,\omega,\cdot}$.

Observe moreover that if $N/T \in (n, n+1)$ then

$$Z^{\mathrm{c}}_{N,\omega,\beta} \geq Z^{\mathrm{c}}_{nT,\omega,\beta} Z^{\mathrm{c}}_{N-nT,\omega,\beta} \geq Z^{\mathrm{c}}_{nT,\omega,\beta} \min_{k=1,\ldots,T-1} Z^{\mathrm{c}}_{k,\omega,\beta} =: c Z^{\mathrm{c}}_{nT,\omega,\beta},$$

(1.55)

and of course $c > 0$. An analogous estimate yields

$$Z^{\mathrm{c}}_{N,\omega,\beta} < \frac{1}{c} Z^{\mathrm{c}}_{(n+1)T,\omega,\beta}.$$

(1.56)

In particular $Z^{\mathrm{c}}_{N,\omega,\beta} \overset{N\to\infty}{\asymp} Z^{\mathrm{c}}_{\lfloor N/T\rfloor T,\omega,\beta}$ and the existence of the free energy limit is established.

The invariance under translation is established in the same way, in the sense that the partition function of the polymer spanning from 1 to $nT+1$ is easily compared from below with the polymer spanning from T to nT and from above with the one ranging from 0 to $(n+1)T$, namely there exists a constant $c \in (0,1]$ such that

$$c Z^{\mathrm{c}}_{(n-1)T,\omega,\beta} \leq Z^{\mathrm{c}}_{nT,\theta\omega,\beta} \leq c^{-1} Z^{\mathrm{c}}_{(n+1)T,\omega,\beta},$$

(1.57)

and this proves that $\mathrm{F}_{\theta\omega}(\beta) = \mathrm{F}_\omega(\beta)$.

The last invariance property is an immediate consequence of the exchangeability of independent variables, so that $\mathbf{P}(\tau_i = t_i, i = 1, \ldots, n)$ is equal to $\mathbf{P}(\tau_i = t_{n-i+1}, i = 1, \ldots, n)$ for any choice of $n \in \mathbb{N}$ and any $t_i \in \mathbb{N}$ such that $t_1 < \ldots < t_n$. $\square$

Beyond general questions, like showing the existence of the free energy, it is more interesting to consider a different parametrization of the problem. For $h \in \mathbb{R}$ we set

$$Z^{\mathrm{c}}_{N,\omega,\beta,h} := \mathbf{E}\left[\exp\left(\sum_{n\in\tau, n\leq N}(\beta\omega_n - h)\right); N \in \tau\right],$$

(1.58)

and ω is a centered periodic sequence. Except in the case $\beta = 0$, this is just a change of parameters. Of course we define

$$\mathrm{F}_\omega(\beta, h) := \lim_{N \to \infty} \frac{1}{N} \log Z^{\mathrm{c}}_{N,\omega,\beta,h}. \tag{1.59}$$

It is once again natural to set $\mathcal{L} := \{(\beta, h) : \mathrm{F}_\omega(\beta, h) > 0\}$ and $\mathcal{D} := \{(\beta, h) : \mathrm{F}_\omega(\beta, h) = 0\}$. As before, it is immediate to see that $\mathrm{F}_\omega(\beta, \cdot)$ is convex and decreasing. Therefore for every β we can set $h_c(\beta) = \sup\{h : \mathrm{F}_\omega(\beta, h) > 0\}$ and $\mathcal{L} := \{(\beta, h) : h < h_c(\beta)\}$. By easy comparison arguments, that is by replacing ω with the constant configuration taking value either $\min_n \omega_n$ or $\max_n \omega_n$, one sees that $h_c(\beta)$ is finite for every β.

We will see in Chapter 3 that this decomposition of the parameter space does correspond to sharply different path behaviors. Of course other questions like the dependence of $h_c(\cdot)$ on ω or the order of the transition are very relevant. For example: what is the value of $h_c(\beta)$ if $\omega_n = (-1)^n$? What is clear is that $h_c(0) = \log \Sigma_K$, since for $\beta = 0$ we are just dealing with a homogeneous pinning model. Moreover, since $-\omega = \theta\omega$, $h_c(\beta) = h_c(-\beta)$. We will give an exact expression for $h_c(\beta)$, we just anticipate the fact $h_c(\beta) > h_c(0)$ for every $\beta > 0$, since this fact gives a first glimpse on the way inhomogeneous environments may favor localization.

1.6.2 *Copolymers, periodic copolymers and selective interfaces*

While a quantitative analysis of the periodic pinning model of Section 1.6.1 is not immediate, on a qualitative level the mechanism of the transition is not new with respect to the corresponding homogeneous model. Now instead we are going to present a case in which the inhomogeneous character of the charge distribution is at the base of the localization mechanism.

An informal introduction to copolymers has been given in Section 1.1. Let us now write explicitly a model:

$$\frac{d\mathbf{P}^a_{N,\omega,\lambda,h}}{d\mathbf{P}}(S) = \frac{1_{\Omega^a_S(N)}}{\widetilde{Z}^a_{N,\omega,\lambda,h}} \exp\left(\lambda \sum_{n=1}^N (\omega_n + h)\, \mathrm{sign}\,((S_{n-1}, S_n))\right), \tag{1.60}$$

where S is a (p, q)-walk and ω is still a sequence of real numbers called *charges* (notice however that it plays a different role than in (1.52)). The sign of the bond (S_{n-1}, S_n) (the nth-monomer) is equal to $+1$, respectively

-1, if the bond lies in the upper half-plane or on the axis, respectively in the lower half-plane, see Figure 1.11. Of course $\Omega^a_S(N)$ is Ω_S if $a = \mathtt{f}$ and it is equal to $\{S : S_N = 0\}$ if $a = \mathtt{c}$: we will for now focus on the latter case, also (and again) because, at least from the free energy viewpoint, the two cases are equivalent. In the case of copolymers we are not going to exclude the simple random walk case $((1,0)$-walk$)$, since most of the literature on copolymers deals with this case: of course if $a = \mathtt{c}$ we need to choose $N \in 2\mathbb{N}$.

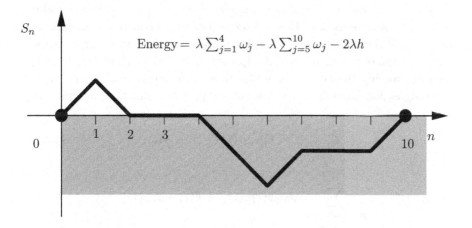

Fig. 1.11 A trajectory of the copolymer. Note that, with our convention, monomers on the interface contribute energetically as if they were in the upper half-plane. This convention is just local and it will be modified later on.

Definition 1.9 covers directly the case of copolymers and without loss of generality it is natural to assume that a periodic copolymer is a copolymer with a centered periodic charge sequence. Note that in this case if $h \geq 0$, by restricting the attention only to S trajectories that are positive in $\{1, \ldots, nT - 1\}$, we have

$$\widetilde{Z}^{\mathtt{c}}_{nT,\omega,\lambda,h} \geq \frac{1}{2} K(nT) \exp(\lambda h nT), \qquad (1.61)$$

so that, if the Laplace asymptotic limit exists, then it leads to a free energy that is greater or equal to λh. We prefer to modify, in a way that is just cosmetic, the Hamiltonian of the system and note that

$$\frac{d\mathbf{P}^a_{N,\omega,\lambda,h}}{d\mathbf{P}}(S) = \frac{1}{Z^a_{N,\omega,\lambda,h}} \exp\left(-2\lambda \sum_{n=1}^{N} (\omega_n + h) \Delta_n\right) \mathbf{1}_{\Omega^a_S(N)}, \quad (1.62)$$

with $\Delta_n := (1 - \text{sign}((S_{n-1}, S_n)))/2$, so that $\Delta_n = 1$ (respectively $\Delta_n = 0$) if the nth–monomer is in the lower half–plane (respectively in the upper half–plane or at the interface). Note that $\widetilde{Z}^a_{N,\omega,\lambda,h} = \exp\left(\lambda \sum_{n=1}^{N} (\omega_n + h)\right) Z^a_{N,\omega,\lambda,h}$ so that $\widetilde{Z}^a_{N,\omega,\lambda,h}/Z^a_{N,\omega,\lambda,h} \overset{N\to\infty}{\asymp} \exp(\lambda h N)$.

It is important to remark that in reality also the copolymer is essentially just a return time model. This is because once we know the returns of S, there is still the uncertainty of the sign of the excursion. However randomizing the sign of the excursions of a (p,q)-walk by independent coin tossing does not modify the law of the walk. So we can integrate with respect to these signs, for example at the level of the partition function and write it as

$$Z^a_{N,\omega,\lambda,h} = \mathbf{E}\left[\exp\left(\mathcal{H}_{N,\omega}(\tau)\right); \Omega^a_S\right], \quad (1.63)$$

with

$$\mathcal{H}_{N,\omega}(\tau) := \sum_{\substack{j=1,\ldots,\mathcal{N}_N(\tau): \\ \tau_j - \tau_{j-1} \geq 2}} \varphi_\lambda \left(\omega(\tau_{j-1}, \tau_j] + h(\tau_j - \tau_{j-1})\right)$$

$$+ \varphi_\lambda \left(\omega(\tau_{\mathcal{N}_N(\tau)}, N] + h\left(N - \tau_{\mathcal{N}_N(\tau)}\right)\right), \quad (1.64)$$

where we have introduced the notations

$$\omega(j,k] := \begin{cases} \sum_{n=j+1}^{k} \omega_n & \text{if } j < k, \\ 0 & \text{if } j = k \end{cases} \quad (1.65)$$

and $\varphi_\lambda(t) = \log\left((1 - \exp(-2\lambda t))/2\right)$. Note that if $\mathcal{N}_N(\tau) = 0$ then the first term in the right-hand side of (1.64) is zero and if $\tau_{\mathcal{N}_N(\tau)} = N$ the last term is zero.

Remark 1.11 *The choice of $+1$ for the sign of a monomer that lies on the interface is clearly asymmetric and it may look unnatural and possibly 0 may be a more reasonable choice. This point is not secondary since in principle one would like to separate the case of neutral interfaces and the case in which there are rewards and/or penalties also for crossing or for lying on the interface, see in particular Section 1.6.3 below. Observe also that one can of course extend the first sum in the right-hand side of (1.64)* *by taking away the condition $\tau_j - \tau_{j-1} \geq 2$ and add an extra term accounting for the change: so the second sum may be viewed as an inhomogeneous pinning term. One way to make this second sum disappear is to decide by flipping a coin the value, 0 or 1, of Δ_n also when $S_{n-1} = S_n = 0$ and this is possibly the case in which the interface is really neutral.*

Without loss of generality, from now on we always assume for copolymers that $\lambda \geq 0$ and that $h \geq 0$.

We have the following strict analog of Proposition (1.10), which is proven in the very same way.

Proposition 1.12 *If $\{\omega_n\}_n$ is a (centered) periodic sequence of charges then the limit of the sequence $\left\{ N^{-1} \log Z^{\mathrm{c}}_{N,\omega,\lambda,h} \right\}_N$ exists and we denote it $\mathrm{F}_\omega(\lambda, h)$. $\mathrm{F}_\omega(\cdot, \cdot)$ takes values in $[0, \infty)$ and it is separately convex in both variables. Moreover $\mathrm{F}_{\theta\omega}(\lambda, h) = \mathrm{F}_\omega(\lambda, h)$ and $\mathrm{F}_\omega(\lambda, h) = \mathrm{F}_{\widetilde{\omega}}(\lambda, h)$.*

Remark 1.13 *Note that $\mathrm{F}_\omega(\cdot, \cdot)$ is separately convex and, in general, not convex. This is simply due to the parametrization we have chosen. By replacing h with h/λ, with obvious interpretation of this change of variables when $\lambda = 0$, we see that the approximating sequence is a sequence of convex functions in these new variables. Therefore if we set $\widetilde{\mathrm{F}}_\omega(\lambda, h) := \mathrm{F}_\omega(\lambda, h/\lambda)$, then $\widetilde{\mathrm{F}}_\omega(\cdot, \cdot)$ is convex.*

Remark 1.14 *Note that $\mathrm{F}_\omega(\cdot, \cdot)$ denotes both the free energy of an inhomogeneous pinning model or the free energy of a copolymer. This notation will be kept also for the disordered model, see below, except that the subscript will disappear in that case. It will be in any case clear from the context if $\mathrm{F}_\omega(\cdot, \cdot)$ refers to the pinning model or to the copolymer. On the other hand, $\mathrm{F}(\cdot)$ (function of one variable) is always the free energy of the homogeneous pinning model.*

The phase diagram of periodic copolymers will be studied in Chapter 3: we suggest the reader Figure 3.3. Note in particular that, excluding the degenerate cases $T = 1$ for general walks ($\omega_n = 0$ for every n) or $T = 2$ for the $(1,0)$-walk (since the energy associated to an excursion is 0), for $h = 0$ the copolymer is localized at an arbitrary small value of λ, a result that at first may look surprising.

1.6.3   *Copolymers with adsorption*

It is very natural to consider interfaces that are not neutral and this corresponds to superimposing the copolymer interaction with the solvents and the pinning interaction at the interface. We are going to focus on the following model, that we call *copolymer model with adsorption*:

$$\frac{d\mathbf{P}^a_{N,\underline{v},\omega}}{d\mathbf{P}}(S) = \frac{1_{\Omega^a_S(N)}}{Z^a_{N,\omega,\underline{v}}} \exp\left(-2\lambda \sum_{n=1}^{N} (\hat{\omega}_n + h)\, \Delta_n + \sum_{n=1}^{N} \left(\beta\omega_n - \widetilde{h}\right) \delta_n \right),$$
(1.66)

with S a (p,q)-walk, $\underline{v} := (\lambda, h, \beta, \widetilde{h})$, $\delta_n = 1_{S_n=0}$ and note that we have distinguished the copolymer charges $\hat{\omega}$ from the pinning ones ω. We will do so only when both pinning and copolymer interactions appear at the same time in the model and in this case ω denotes $\{(\hat{\omega}_n, \omega_n)\}_n$, otherwise, that is when either $\lambda = 0$ or $\beta = \widetilde{h} = 0$, $\omega = \{\omega_n\}_n$ as usual.

The straightforward generalization of Proposition 1.10 and of Proposition 1.12 holds. So we can speak of a free energy $\mathtt{F}_\omega(\underline{v})$ at $\underline{v}$ and this free energy is non–negative. As we shall see in Chapter 3 the phase diagram of the periodic copolymer with adsorption is richer that the copolymer phase diagram or the pinning phase diagram, beyond the fact that it is four–dimensional.

Remark 1.15   *One could consider a model of copolymer with adsorption which is more general than the one given in (1.66). We are in fact rewarding, or penalizing, the passage by the interface, and, indirectly, also staying on the interface. But we could give a reward (or a penalization) when the monomer is at the interface, and not, or not only, when it crosses the interface. This corresponds to introducing an interaction that depends on $\bar{\delta}_n := 1_{S_{n-1}=S_n=0}$. We will consider in some cases this type of interaction, but not systematically. Note in particular that the interaction suggested as a possible alternative definition of Δ_n in Remark 1.11 corresponds to adding $\lambda \sum_{n=1}^{N} \hat{\omega}_n \bar{\delta}_n$ to the energy of the copolymer. Once again we face the problem of deciding whether the interface is neutral or not, an issue further discussed in Chapter 3 and Chapter 6.*

1.7 Fully Inhomogeneous Charge Distributions and Disordered Polymer Models

Beyond certain specific applications, the *regularity*, clearly present in periodic models, is a strong limitation. In particular, weakly inhomogeneous models can be reduced to homogeneous models by *coarse graining* on a the finite length scale $T(\omega)$. One could for example *decimate* the model on this scale: this means taking the marginals of the polymer measure on $\{S_{nT(\omega)}\}_{n=0,1,\ldots}$. It is a lengthy (if $T(\omega) > 2$), but instructive exercise to try to describe in detail the homogeneous process arising after such a procedure. The problem is essentially connected to the fact that the decimated free polymer $\{S_{nT(\omega)}\}_{n=0,1,\ldots}$ under $\mathbf{P}$ may cross the interface without touching it and to each typology of crossing corresponds to a different energy contribution (and not necessarily of the same sign). In principle one could think of tackling such a problem via decimation, but it appears to be easier to attack the model directly and this is what we do in Chapter 3. This coarse graining idea however suggests that the arising phenomena may be reduced to phenomena that are observable in homogeneous pinning models: this is only partially true, in the sense that it becomes true if one considers a class of pinning models that is greatly generalized with respect to the one we propose here. And in fact, new phenomena arise, see Chapter 3.

Out of the class of weakly inhomogeneous charge distributions, the application of coarse graining ideas becomes much less evident.

The first, and central for this work, example of fully inhomogeneous charge distribution comes from choosing ω as typical configuration of a family of random variables.

The definition of the polymer measures $\mathbf{P}_{N,\omega,\beta}^a$, $\mathbf{P}_{N,\omega,\lambda,h}^a$ and $\mathbf{P}_{N,\omega,\underline{v}}^a$ is as in Section 1.6, and we simply write $\mathbf{P}_{N,\omega}^a$. Of course now $\mathbf{P}_{N,\omega}^a$ may be looked upon as a standard probability measure over Ω_S, or as a random variable that takes values in the space of probability measures over Ω_S. Let us therefore make more precise the definition of ω. First of all $\omega := \{(\hat{\omega}_n, \omega_n)\}_n$ is an element of Ω, that with the σ–algebra $\mathcal{A}$ forms a measurable space on which we put the measure $\mathbb{P}$. Therefore the canonical projections $\omega \mapsto \hat{\omega}_n$ and $\omega \mapsto \omega_n$, both maps from Ω to $\mathbb{R}$, are random variables and $\hat{\omega}_n$ and ω_n are realizations of these random variables. However introducing specific notations to distinguish between random variables and realizations is a bit cumbersome and, in our case, superfluous, so we will never do that.

We will consider various $\mathbb{P}$, so various laws of ω, but we single out a typical hypothesis:

Definition 1.16   We will say that the family of random variables ω is IID, if the two sequences $\{\hat{\omega}_n\}_n$ and $\{\omega_n\}_n$ are IID (but in general $\hat{\omega}_1 \not\sim \omega_1$, recall that $\sim$ denotes equality in law) and the whole family ω is a family of independent variables. We say that ω_1 is exponentially integrable, if

$$\mathrm{M}(t; \omega_1) := \mathbb{E}\left[\exp\left(t\omega_1\right)\right] < \infty, \tag{1.67}$$

for every $t \in \mathbb{R}$. If (1.67) holds only for t in a neighborhood of 0 we say that ω_1 is locally exponentially integrable. We call $D_{\mathrm{M}(\cdot;\omega_1)}$, shortened to D_{M} whenever possible, the interval on which $\mathrm{M}(\cdot)$ is finite. Of course the same definitions hold for $\hat{\omega}_1$. The family of random ω variable is (locally) exponentially integrable if all variables are (locally) exponential integrable.

Definition 1.17   *Standard assumptions on ω.* Unless specified otherwise, if ω is random it has to be meant IID and locally exponentially integrable. Moreover, without loss of generality, ω_1 and $\hat{\omega}_1$ are centered and $\mathbb{E}[\omega_1^2] = \mathbb{E}[\hat{\omega}_1^2] = 1$.

A very important point is that our main interest is on $\mathbf{P}_{N,\omega}^a$ in the sense of a *quenched model*, that is we look for results that hold for typical realizations of ω (most of them will be $\mathbb{P}(\,\mathrm{d}\omega)$-a.s. results). So, in principle, there is no need to define Ω and S (or τ) on the same probability space. In practice, it turns out to be practical to work on $(\Omega \times \Omega_S, \mathcal{A} \otimes \mathcal{F}, \mathbb{P} \otimes \mathbf{P})$, that is with ω independent of S (or τ).

We will use the term *disordered* as a synonym with *quenched*. In particular the quenched charges ω will be simply called *disorder*.

Of course this time $Z_{N,\omega}^a$ may be looked upon as a random variable and, in any case, one has to specify for which ω the limit defining the quenched free energy

$$\mathrm{F} = \lim_{N\to\infty} \frac{1}{N} \log Z_{N,\omega}^a, \tag{1.68}$$

exists. We will show the existence of such a limit both $\mathbb{P}(\,\mathrm{d}\omega)$-a.s. and in the $L^1(\Omega, \mathcal{A}, \mathbb{P})$ sense, even under assumptions on ω that are much weaker than the standard ones. And we stress that F depends on the parameters, β, λ, etc..., but not on $a = \mathtt{c}$ or $\mathtt{f}$ nor on the particular realization ω: the latter property goes under the name of *self-averaging property* (of the free energy). Therefore

$$\mathrm{F} = \lim_{N \to \infty} \frac{1}{N} \mathbb{E} \log Z^a_{N,\omega}, \tag{1.69}$$

and the quantity on the right-hand side is called *quenched averaged free energy*. So, quenched free energy and quenched averaged free energy coincide.

Notice that, exactly as in the periodic set-up,

$$Z^c_{N,\omega,\underline{v}} \geq \frac{K(N)}{2} \exp\left(\beta\omega_N - \widetilde{h}\right), \tag{1.70}$$

so that we see that $\mathrm{F} \geq 0$ and we are led to partitioning the parameter space into the two sets $\mathcal{D}$ and $\mathcal{L}$. Since it is quite crucial throughout this work, let us spell it out explicitly.

Definition 1.18 If $\underline{v}$ is an element of the parameter set and $\mathrm{F}(\underline{v})$ is the free energy at $\underline{v}$, then we partition the parameter set into

$$\mathcal{L} := \{\underline{v} : \mathrm{F}(\underline{v}) > 0\} \quad \text{and} \quad \mathcal{D} := \{\underline{v} : \mathrm{F}(\underline{v}) = 0\}. \tag{1.71}$$

If $\underline{v} \in \mathcal{L}$, respectively $\underline{v} \in \mathcal{D}$, then we say that the system is in the localized regime, respectively the delocalized regime.

Characterizing the phase diagram, as well as characterizing the path properties of $\mathbf{P}^a_{N,\omega}$ in the limit of large N, will be at the heart of our work.

In order to have an idea of the (serious) complications and of the challenging questions that naturally arise in a disordered context let us, for example, consider the disordered copolymer model, with $\omega_1 \sim \mathcal{N}(0,1)$, that is ω_1 is a standard Gaussian random variable. Let us observe that by Jensen inequality

$$
\begin{aligned}
\mathbb{E} \log Z^f_{N,\omega,\lambda,h} &\leq \log \mathbb{E} Z^f_{N,\omega,\lambda,h} \\
&= \log \mathbb{E}\mathbb{E}\left[\exp\left(-2\lambda \sum_{n=1}^{N} (\omega_n + h)\Delta_n\right)\right] \\
&= \log \mathbb{E}\left[\prod_{n=1}^{N} \mathrm{M}(-2\lambda\Delta_n)\exp(-2\lambda h\Delta_n)\right] \\
&= \log \mathbb{E}\left[\exp\left(\widetilde{\beta}\sum_{n=1}^{N}\Delta_n\right)\right],
\end{aligned}
\tag{1.72}
$$

where $\mathrm{M}(\cdot) := \mathrm{M}(\cdot; \omega_1)$ and we have of course used the Fubini–Tonelli Theorem in the second step and the IID property of ω in the third step. In the last term $\widetilde{\beta} = \log \mathrm{M}(-2\lambda) - 2\lambda h$. Since $\mathrm{M}(-2\lambda) = \exp(2\lambda^2)$, if $h \geq \lambda$ then $\widetilde{\beta} \leq 0$, so that $\mathbb{E}\log Z^{\mathrm{f}}_{N,\omega,\lambda,h} \leq 0$. Since we know that $\mathrm{F}(\lambda, h) \geq 0$, $\mathrm{F}(\lambda, h) = 0$ for $h \geq \lambda$ and the polymer is delocalized. This means that h wins over the localization effect due to the change in solvent at the interface, and one certainly expects, in analogy with the transient behavior of the homogeneous polymers in the delocalized regime, that the polymer does not come back to the interface. However in a typical disorder configuration ω one finds arbitrarily long *atypical* stretches: for example, for every $L \in \mathbb{N}$ one can find a $n(\omega) \in \mathbb{N}$ such that $\omega_n \leq -2h$ for $n = n(\omega)+1, \ldots, n(\omega)+L$. Such a region of course strongly penalizes delocalized trajectories: this phenomenon is not there for homogeneous or weakly inhomogeneous models and therefore this casts some doubts about transferring the intuition that we have developed for homogeneous or weakly inhomogeneous models to disordered models.

Remark 1.19 *By writing the analog of formula (1.8) (or (1.29)) for inhomogeneous models, it is clear that if $\Sigma_K < 1$, that is if the underlying process is transient (or defective), the model $\mathbf{P}^{\mathrm{c}}_{N,\underline{v},\omega}$ is equivalent to a model in which $K(\cdot)$ is replaced by $K(\cdot)/\Sigma_K$ (recurrent!) and to which a homogeneous pinning interaction, in fact a penalization, of $\log \Sigma_K$ is added. This means that $\widetilde{h}$ is replaced by $\widetilde{h} - \log \Sigma_K$. The situation is just about the same for $\mathbf{P}^{\mathrm{f}}_{N,\underline{v},\omega}$, where the procedure we propose changes the measure and the partition function, but in a way that is totally negligible for large N (the difference is just connected to the pinning penalization given also to the last incomplete excursion). This remark allows to simplify, or at least to shorten, several proofs.*

Remark 1.20 *Is the process S, distributed according to $\mathbf{P}^a_{N,\underline{v}}$, still a Markov process? The analogous question in the case of models built on τ is: does the renewal property or, rather, a generalization of it (that takes into account the inhomogeneous character of our framework) still hold? The answer is positive, as it becomes clear if we take a Gibbs measure view point [Georgii (1988)], since the interacting potentials are only of 1–body type. This fact is however absolutely elementary: for example if $n \in \{1, \ldots, N-1\}$, $A \subset \{1, \ldots, n\}$ and $B \in \{n+1, \ldots, N\}$ then by playing with* *ratios of partition functions we see that*

$$\mathbf{P}^{\mathrm{f}}_{N,\omega}\left(A \cup \{n\} \cup B \subset \tau\right) = \mathbf{P}^{\mathrm{f}}_{N,\omega}\left(A \cup \{n\} \subset \tau\right) \mathbf{P}^{\mathrm{f}}_{N-n,\theta^n\omega}\left(\theta^{-n}B \subset \tau\right).$$

$$(1.73)$$

This is clearly a (generalized) renewal property. One can easily generalize this formula and obtain analogous formulas for S based models. We will repeatedly use this property, but mainly in a non-explicit way and we will mostly manipulate (restricted) partition functions. Homogeneous Markov and renewal processes are manageable due to their local nature, but, in inhomogeneous frameworks, they may instead display sharply nonlocal features and tools to analyze them go well beyond the tools used for homogeneous systems.

1.7.1 Annealed models

Taking a quenched approach means treating the two random *ingredients* of the model on very different grounds. This is natural since we imagine that the charge is associated once and for all to a monomer, think for example of the Poland–Scheraga case: the sequence of bases in a stretch of DNA is given, and the polymer instead fluctuates in time (even if we focus on equilibrium properties).

But we could as well consider the case in which ω and S (or τ) are considered on the same ground. This corresponds to looking at the *annealed* model (say, in the copolymer case)

$$\frac{\mathrm{d}(\mathbb{P} \otimes \mathbf{P})^{\mathrm{f}}_{N,\lambda,h}}{\mathrm{d}(\mathbb{P} \otimes \mathbf{P})}(\omega, S) = \frac{1}{Z^{\mathrm{f}}_{N,\lambda,h}} \exp\left(-2\lambda \sum_{n=1}^{N} (\omega_n + h)\,\Delta_n\right).$$

$$(1.74)$$

Note that $Z^{\mathrm{f}}_{N,\lambda,h}$ is exactly the quantity that appears in (1.72) after having applied Jensen inequality. Notice moreover that after the exchange in the order of integration $\mathbb{EE} \to \mathbf{EE}$ and explicit computation of the Boltzmann factor one gets to the partition function of a homogeneous model, in the specific case a penetrable substrate model, that is the model with energy $\widetilde{\beta} \sum_{n=1}^{N} \Delta_n$, see (1.72). Throughout the text, by *annealed model* we will mostly mean this homogeneous model, which is in reality a marginal of the annealed model defined in (1.74). Such a model is solvable, it is a particular case of the general model considered in Section 1.3.1, and we have seen that the trajectories are in any case delocalized (above or below the axis, according to the sign of $\widetilde{\beta}$).

But if one looks at the trajectory of the couple (ω, S) it is clear that this delocalization phenomenon is due to the charges that rearrange themselves in order to favor the corresponding delocalized behavior (after all, the free polymer is delocalized). The extreme case is when $h = \lambda$, so that $\tilde{\beta} = 0$ and S is absolutely not disturbed by the presence of the charges, that evidently rearrange themselves in a rather subtle way not to disturb the S trajectory at all. In order to make these statements precise it is sufficient to compute the distribution of ω conditioned by a trajectory of S. We will not go into the details of this curious phenomenon, which is however due to the very different nature of ω and S: $\{\Delta_n\}_n$ is of course much more *rigid*, in the sense of difficult to flip, than ω. And the same is true for $\{\delta_n\}_n$ in the pinning case, so that the mechanism behind the annealed versions of the models we consider turns out to be rather elementary.

All the same, the simple steps in (1.72), and the analogous ones in more general cases, lead to important estimates: the *annealed bounds*, that say that the quenched free energy is smaller or at most equal to the annealed free energy (some ideas on how to go beyond this bound are explained in Section 5.3).

1.8   On the Return Time Viewpoint

It is by now clear that the models that we are considering may be reduced, completely or partially, to return time models, when they are not defined from the start simply via return times: in particular, there is no loss of information with this new viewpoint if one just considers the partition function of the model.

Let us write explicitly the general return time model: this will be useful in Chapter 4, where we are going to prove the existence of the free energy, but also on a conceptual level, since it suggests a general abstract class of models to which the techniques we develop may be applied. In what follows the reader should keep in mind in particular formulas (1.63) and (1.64).

For every typical realization of a renewal process τ we introduce the general Hamiltonian

$$\mathcal{H}_{N,\omega}(\tau) := \sum_{j=1}^{\mathcal{N}_N(\tau)} \mathcal{H}_\omega^{\mathrm{exc}}(\tau_{j-1}, \tau_j) + \mathcal{H}_\omega^{\mathrm{i-exc}}(\tau_{\mathcal{N}_N(\tau)}, N), \qquad (1.75)$$

where

$$\mathcal{H}_\omega^{\mathrm{exc}}(i, k) := \varphi_{(1)}(\omega_k) + \varphi_{(2)}(\hat{\omega}(i, k]) + \varphi_{(3)}(\hat{\omega}_k) \mathbf{1}_{k-i=1}, \qquad (1.76)$$

is defined for $i < k$ and the three functions in the right-hand side are continuous functions with sub-linear growth at infinity, that is $\sup_{|t|>1} |\varphi_{(i)}(t)/t| < \infty$, $i = 1, 2, 3$. Moreover

$$\mathcal{H}_\omega^{\text{i-exc}}(i, k) := \varphi_{(2)}(\hat{\omega}(i, k]), \tag{1.77}$$

for every $i \le k$ and we insist on $\varphi_{(2)}(0) = 0$ so this term, that is the energetic contribution of the last incomplete excursion, is zero if $j = k$. We will also make the assumption that $\varphi_{(2)}(\cdot)$ is bounded below and that there exists a constant c such that

$$\varphi_{(2)}(t_1 + t_2) \le c + \varphi_{(2)}(t_1) + \varphi_{(2)}(t_2), \quad \text{for every } t_1, t_2 \in \mathbb{R}. \tag{1.78}$$

In the examples we have treated $\varphi_{(1)}(\cdot)$ is an affine function and $\varphi_{(2)}(\cdot)$ can be read out of (1.64). The function $\varphi_{(3)}(\cdot)$ is not present in (1.64), but it would be present with a different choice of sign$((0, 0))$ and, in general, it is an affine function. With reference to the same formula, $\varphi_{(2)}(\cdot) = \varphi_\lambda(\cdot)$ and $c = \log 2$ in (1.78). A list of properties of $\mathcal{H}_{N,\omega}(\tau)$ is found in Chapter 4.

1.9 On Related Classes of Models

In this section we give a very quick and limited overview of what we left out. We will essentially talk only of two issues:

(1) Going beyond the framework given by Definition 1.4, that is relaxing the conditions on the return distribution $K(\cdot)$.
(2) Giving rewards (or penalties) on structures more general than a line.

An even quicker (due) discussion on self-avoiding models is instead relegated to the bibliographic complements (Section 1.10).

1.9.1 *More general return times*

Definition 1.4, that is *almost polynomial decay*, has been chosen because it gives a practical and intuitive working framework that catches several of the interesting and relevant models.

However here we point out that:

- A limitation of Definition 1.4 comes from asking for a precise decay. In reality many of the results we present depend only on bounds on

the decay, like for example the upper or the lower bound in (1.25), and
some even only along a subsequence.

- It is absolutely natural to choose $K(\cdot)$ with sub-exponential, but not
  polynomial, decay, like for example $K(n) \overset{n \to \infty}{\sim} \exp(-n^c)$, $c \in (0,1)$.
  These cases are implicitly treated in great generality, but there are arguments for which a polynomial type decay is crucial (in particular,
  the *rare stretch* arguments of Chapter 5 and Chapter 6). On the other
  hand, models with super-exponential return distributions, for example
  $c > 1$ in the formula above, can be tackled with the techniques we are
  using, but they have certain characteristics that make them of limited
  interest from our viewpoint. For example, if $K(\cdot)$ is a probability, the
  interacting potentials we consider cannot induce a transition. Added to
  that, the argument that leads to the non-negativity of the free energy
  does not apply and in fact the free energy, for example in the homogeneous pinning, can take any real value. It is also rather easy to exhibit
  models (it is essentially the typical situation) for which leaving free or
  constraining the endpoint leads to substantially different models. Of
  course one may think of having such a super-exponential decay only
  along subsequences, but this is essentially the problem of relaxing the
  condition of positivity on $K(\cdot)$.

- A particular discussion deserves instead the case of exponential return
  times, meaning by this for example return distributions of the type
  $K(n) = \sigma \exp(-\kappa n) n^c$, $\kappa > 0$ and $c \in \mathbb{R}$ and $\sigma > 0$. Once again look
  at the easiest case, homogeneous pinning, and consider the constrained
  endpoint case. Just reconsider (1.8) and it becomes apparent that we
  have

$$Z_{N,\beta}^{\mathrm{c}} = \exp(-\kappa N) \sum_{n=1}^{N} \sum_{\substack{\ell \in \mathbb{N}^n: \\ \sum_{j=1}^{n} \ell_j = N}} \prod_{j=1}^{n} \exp(\beta) \sigma \ell_j^{c}. \qquad (1.79)$$

Notice in particular that there is no need to ask for $\kappa > 0$, but of course
one has to give up the interpretation of $K(\cdot)$ as a sub-probability (it
would rather be a combinatorial term, see in particular the way the
Poland–Scheraga model has been introduced). Regardless of the value
of κ, we realize that, if $c < -1$, then $\exp(\kappa N) Z_{N,\beta}^{\mathrm{c}}$ is the partition
function of the homogeneous pinning model with polynomial decay of
the return times (and critical point β_c). Therefore there is a transition,
between free energy equal to $-\kappa$ to larger than $-\kappa$, at β_c. And it is

not difficult to see that it is the *standard* localization–delocalization transition. If instead $c \geq -1$ then $\mathrm{F}(\beta) > -\kappa$ for every β and there is no transition (in fact the model is always localized). A proof of this fact follows simply by observing that, since $\sum_n \sigma n^c = \infty$, for every $\beta \in \mathbb{R}$ one can find $b > 0$ such that $\sum_n \sigma n^c \exp(-bn) = \exp(-\beta)$, and the free energy of the system is $-\kappa + b$. In words, what is happening if $c \geq -1$ is that even if the returns are penalized, $\beta < 0$, the polymer can gain entropically by making long returns (the smaller β, the longer the returns). We insist that these arguments do not require (at all) that $n \mapsto \sigma \exp(-\kappa n)n^c$ is a probability density, and this makes clear that the model is drastically dependent on boundary conditions. Suppose in fact that $\sum_n \sigma \exp(-\kappa n)n^c =: p < 1$, then $Z^{\mathrm{f}}_{N,\beta} \geq \mathbf{P}(\tau \cap \{1, \dots, N\}) \geq (1 - p)$. Therefore the free energy of the free model is non-negative: it is even more intuitive the fact that if $\beta \leq 0$ this model is delocalized, so free and constrained endpoint models have little to do with each other. The phenomenon we have described now becomes somewhat enlightening when the underlying renewal is given by the returns of a one dimensional lattice random walk with positive expectation increments. In this case the walk is transient, but constraining the endpoint leads to a centered random walk, by exchangeability of the increments.

1.9.2 *Rewards and penalties on general structures*

We introduced and, as a matter of fact, we will consider only polymer models in which the interaction is on a line or on a half space. It is of course very natural to consider the same phenomenon on more general structures. A prototypical example is the following: let S be the walk introduced in Section 1.3, that is a walk with increments in $\mathbb{Z}$, and let s be a function from $\mathbb{N}$ to $\mathbb{Z}$. Consider then the interaction energy $\beta \sum_{n=1}^{N} \mathbf{1}_{S_n = s(n)}$. We have seen that if $s(\cdot)$ is constant, S will stick to $s(\cdot)$ as soon as $\beta > 0$. It is on the other hand not difficult to show for example that if $s(n) = \lfloor an \rfloor$, $a \neq 0$, for every n, the walk will stick to the line only if β is sufficiently large. But of course one can consider much more general functions: for example it can be shown [D. Ioffe, private communication (2005)] that if $s(\cdot)$ is a typical configuration of a centered random walk with finite variance (we are dealing thus with a quenched model) then S will localize for any $\beta > 0$.

This new class of problems naturally leads toward a number of other (polymer, random walk, diffusion) models that we will not consider, notably the well known problem of *directed polymers in random environment* (see *e.g.* [Comets *et al.* (2004)]) or the problems of diffusion among random obstacles [Sznitman (1998)], in which localization phenomena naturally arise. And this if one is restrained only to diffusion models. Some remarks will be made here and there along the development of the next chapters, but, of course, a line has to be drawn somewhere.

1.10 Bibliographic Complements

Complements on Section 1.2

It is difficult to trace back in the literature who first did these types of computations and when. If we take a mathematician's standpoint, computing the free energy of pinning models is just computing the leading asymptotic behavior of the Laplace transform of the number of renewals up to time N, as N tends to infinity. This certainly dates back to the forties or fifties (we cite [Feller (1966)] and [Feller (1971)] and the several references therein). But of course there is a physical insight, or a physical interpretation, that goes beyond the mere computation and even for this it is difficult to be sharp with references, since the very same computation has been repeated over and over. We choose to refer to the beautiful review [Fisher (1984), in particular Section 6] (we have cited an enlightening paragraph of this work in the preface). We draw the attention of the reader on the fact that the computations in [Fisher (1984), Section 6], as well as in essentially all the physical literature, are different from the ones we present since they aim at computing the radius of convergence of the series $\sum_N Z_{N,\beta}^{\mathrm{c}} z^N$. Since $Z_{N,\beta}^{\mathrm{c}} \asymp \exp(\mathrm{F}(\beta)N)$, knowing the radius of convergence is equivalent to knowing the free energy. In [Fisher (1984)] and in the physical papers cited therein there are then ingenious arguments to extract more information on the properties of the system. We take instead a *renewal theory* approach ([Feller (1971)], [Asmussen (2003)]) and aim directly for the asymptotic behavior of $Z_{N,\beta}^{\mathrm{c}}$: this yields naturally estimates that go well beyond catching the leading exponential behavior.

Complements on Section 1.3

See Appendix C.

Complements on Section 1.4

The literature on DNA denaturation and Poland–Scheraga models is extremely vast, including various review articles, see in particular [Richard and Guttmann (2004)]. For a more concise, but still rather clear exposition we suggest [Kafri *et al.* (2000)].

Complements on Section 1.5

For force induced unzipping or delocalization in model systems, but motivated by biopolymer experiments, we signal [Marenduzzo *et al.* (2001)] and [Lubensky and D. R. Nelson (2000)], even if our approach is rather different. An example of application of the same models outside of the biopolymer context can be found in [Kafri *et al.* (2006)]. Formula (1.41) extends easily to disordered cases, see [Giacomin (2004), Ex. 2.3] and from it a number of conclusions can be easily drawn.

Complements on Section 1.6

See Chapter 3 and relative bibliographic complements.

Complements on Section 1.7

See Chapters 4 to 9 and relative bibliographic complements. The annealed model is treated for example in [Sinai and Spohn (1996)].

We point out that recently a number of very interesting books on various disordered models have been published: in particular [Bolthausen and Sznitman (2002)], [Bovier (2006)], [Sznitman (1998)] and [Talagrand (2003)].

Complements on Section 1.9

Very general return times are treated in [Alexander and Sidoravicius (2006)], that deals with disordered pinning models: the homogeneous case is treated in detail, but only at the level of the contact fraction (namely, at the level of Large Deviations estimates).

Rigorous results on copolymers based on self-avoiding non-directed walks can be found in [Madras and Whittington (2003)]. There is also a considerable amount of numerical work in this direction, see *e.g.* [Causo and Whittington (2003)].

Another research direction that we will not take into account is the study of non-flat interfaces or of polymers in multi-interface environments. Models for such situations based on directed walks and rigorously analyzed may be found for example in [den Hollander and Wüthrich (2004)], [den Hollander and Whittington (2006)] and [Pétrélis (2006)]

Chapter 2

The Homogeneous Pinning Model

In this chapter we are going to treat in detail the general homogeneous pinning model introduced in Section 1.2.2 by giving its density with respect to the free process, the renewal τ of law $\mathbf{P}$, in equation (1.23) that we recall here:

$$\frac{\mathrm{d}\mathbf{P}^a_{N,\beta}}{\mathrm{d}\mathbf{P}}(\tau) = \frac{1}{Z^a_{N,\beta}} \exp\left(\beta \mathcal{N}_N(\tau)\right) \mathbf{1}_{\Omega^a_S(N)}. \tag{2.1}$$

We will start by studying the order of the transition, thus completing what we have started in Section 1.2.2. But the main focus of this chapter is on showing how the tools of renewal theory yield sharp estimates on the partition function of the model and how from these sharp estimates one can get very precise information on the path properties of the polymer.

2.1 The Free Energy

We have already observed in Section 1.2 that the limit of the sequence $\{(1/N)\log Z^c_{N,\beta}\}_N$ exists. This limit is the free energy, denoted by $\mathrm{F}(\beta)$. The existence of this limit is established with the proof of Proposition 1.1, which yields in reality a substantially stronger result and that also provides an implicit expression for $\mathrm{F}(\beta)$: it is determined by

$$\sum_n K(n) \exp(-\mathrm{F}(\beta)n) = \exp(-\beta), \tag{2.2}$$

if $\sum_n K(n) \geq \exp(-\beta)$, and $\mathrm{F}(\beta) = 0$ otherwise. Moreover $F(\cdot)$ is continuous, non-decreasing, and it is real analytic except at the critical point $\beta_c := \inf\{\beta : \mathrm{F}(\beta) > 0\} = -\log \Sigma_K$ (Figure 1.5). We will now focus on the question: what is the regularity of $\mathrm{F}(\cdot)$ at β_c?

We are going to prove the following:

Theorem 2.1 *For every choice of $\alpha \geq 0$ and $L(\cdot)$, cf. Definition 1.4, there exists a slowly varying function $\hat{L}(\cdot)$ such that for every $\delta > 0$*

$$\mathrm{F}(\beta_c + \delta) = \delta^{1/\min(1,\alpha)} \, \hat{L}(1/\delta). \tag{2.3}$$

In particular

(1) if $\sum_{n\in\mathbb{N}} nK(n) < \infty$ then $\lim_{\delta \searrow 0} \hat{L}(1/\delta) = \Sigma_K / \sum_{n\in\mathbb{N}} nK(n)$;

(2) if $\alpha \in (0,1)$ then $\hat{L}(1/\delta) = (\alpha/\Gamma(1-\alpha))^{1/\alpha}\delta^{-1/\alpha}R_\alpha(\delta)$ and $R_\alpha(\cdot)$ is asymptotically equivalent to the inverse of the map $b \mapsto b^\alpha L(1/b)/\Sigma_K$, cf. property L.5 (Appendix A.4). If $\lim_{N\to\infty} L(N) = c$, then $\hat{L}(1/\delta)$ converges, as $\delta \searrow 0$, to $(\alpha\Sigma_K/(c\Gamma(1-\alpha)))^{1/\alpha}$.

(3) If $\alpha = 0$, (2.3) means that $\mathrm{F}(\delta)$ vanishes faster than any power of δ (see (2.9)).

(4) If $\alpha = 1$ and $m_K = \infty$ then $\hat{L}(\cdot)$ vanishes at infinity.

Proof. By Remark 1.19, there is no loss of generality in assuming $\Sigma_K = 1$ (and therefore $\beta_c = 0$ and $\mathrm{F}(\delta) \searrow 0$ as $\delta \searrow 0$). More precisely, the free energy of a model obtained from the renewal process with inter-arrival law given by the sub-probability $K(\cdot)$ coincides with the one with inter-arrival law given by the probability $K(\cdot)/\Sigma_K$, with argument translated of $-\log\Sigma_K$. In what follows the properties L.x refer to the list in Appendix A.4.

If $m_K < \infty$ we proceed as follows: first observe that, for $b > 0$, $\sum_n K(n)(1-\exp(-bn))/b = \sum_n nK(n)c(bn)$ with $c(x) = (1-\exp(-x))/x$. Notice that $\sup_{x\in(0,\infty)} c(x) < \infty$ and that $\lim_{x\searrow 0} c(x) = 1$, so that, by the Dominated Convergence Theorem, $\sum_n K(n)(1-\exp(-bn))/b \sim m_K$ as $b \searrow 0$. Therefore

$$\mathrm{F}(\delta)m_K \sim 1 - \sum_n \exp(-\mathrm{F}(\delta)n)K(n) = 1 - \exp(-\delta) \sim \delta, \tag{2.4}$$

where the equality is (2.2) and the asymptotic relations are meant in the limit of $\delta \searrow 0$. This proves (2.3), with $\lim_{x\to\infty} \hat{L}(x) = 1/m_K$.

For $\alpha \in (0,1)$ we observe that, since $\sum_{n>N} K(n) \sim L(N)N^{-\alpha}/\alpha$ for N large (property L.3), from Theorem A.2 we get

$$1 - \sum_n \exp(-bn)K(n) \overset{b\searrow 0}{\sim} \frac{1}{\alpha}\Gamma(1-\alpha)b^\alpha L(1/b), \tag{2.5}$$

so that, by (2.2), we obtain

$$\frac{1}{\alpha}\Gamma(1-\alpha)F(\delta)^{\alpha}L(1/F(\delta)) \overset{\delta \searrow 0}{\sim} \delta. \tag{2.6}$$

It is now just a matter of inverting this asymptotic expression by using property L.5.

We treat by hand the marginal cases $\alpha = 0$ and $\alpha = 1$ with $m_K = \infty$: in both cases we use the summation by parts formula (A.25), so that

$$1 - \sum_n \exp(-bn)K(n) \overset{b \searrow 0}{\sim} b \sum_n \overline{K}(n)\exp(-bn). \tag{2.7}$$

If $\alpha = 0$ then (property L.4) $\overline{K}(\cdot)$ is vanishing in a slowly varying way at infinity (in this case define $\overline{K}(\cdot)$ on $(0,\infty)$ by linear interpolation, so that it is a monotonic function). By L.2, $\sum_{n=0}^{N}\overline{K}(n) \sim N\overline{K}(N)$ for N large and therefore, by Theorem A.2, as $b \searrow 0$ we have that

$$b\sum_n \overline{K}(n)\exp(-bn) \sim \overline{K}(1/b). \tag{2.8}$$

By (2.2) we obtain

$$F(\delta) \overset{\delta \searrow 0}{\sim} 1/\overline{K}^{-1}(\delta). \tag{2.9}$$

If instead $\alpha = 1$ and $m_K = \infty$ then, by property L.3, $\overline{K}(N) \sim L(N)/N$ as $N \to \infty$. Keeping into account that $\sum_n \overline{K}(n) = \infty$, we see that $\sum_n \overline{K}(n)\exp(-bn)$ is asymptotically equivalent to $\int_1^\infty (L(r)/r)\exp(-br)\,\mathrm{d}r =: \widetilde{L}(1/b)$. It is now immediate to verify that $\widetilde{L}(x/b)/\widetilde{L}(1/b)$ converges to 1 as $b \searrow 0$ for every $x > 0$, so that $\widetilde{L}(\cdot)$ is slowly varying. Let us now plug these estimates into (2.2) to obtain

$$F(\delta)\widetilde{L}(1/F(\delta)) \overset{\delta \searrow 0}{\sim} \delta, \tag{2.10}$$

and once again the result follows by inverting the asymptotic relation. □

2.2   Renewal Theory and Sharp Estimates

We start by introducing the notation

$$\widetilde{K}_\beta(n) := \begin{cases} \exp(\beta)K(n)\exp(-\mathrm{F}(\beta)n) & \text{if } \beta \geq \beta_c, \\ \exp(\beta)K(n) & \text{if } \beta < \beta_c. \end{cases} \qquad (2.11)$$

Notice that, thanks to (2.2) and the definition of β, $\widetilde{K}_\beta(\cdot)$ is a discrete probability density if $\beta \geq \beta_c$ and, in general, it is a discrete sub-probability density. The law of the associated renewal process will be denoted by $\widetilde{\mathbf{P}}_\beta$: it is positive recurrent if $\beta > \beta_c$ and it is transient if $\beta < \beta_c$. At the critical point the renewal can be either positive or null recurrent, depending on the tail of $K(\cdot)$.

We are going to prove:

Theorem 2.2   *As $N \to \infty$ we have:*

(1) (The localized regime.) If $\beta \in \mathcal{L}$ then

$$Z_{N,\beta}^a \sim C_a \exp(\mathrm{F}(\beta)N), \qquad (2.12)$$

with

$$C_c := \frac{1}{m_{\widetilde{K}_\beta}} \quad and \quad C_f := \frac{1 - \exp(-\beta)}{(1 - \exp(-\mathrm{F}(\beta)))\, m_{\widetilde{K}_\beta}}. \qquad (2.13)$$

(2) (The strictly delocalized regime.) If $\beta \in \overset{\circ}{\mathcal{D}}$ then

$$Z_{N,\beta}^c \sim \frac{\exp(\beta)}{(1-\varrho)^2}\, K(N) \quad and \quad Z_{N,\beta}^f \sim \frac{1}{1-\varrho}\overline{K}(N) + K(\infty)\widetilde{\mathsf{C}}_\beta, \qquad (2.14)$$

where $\varrho := \exp(\beta)\Sigma_K\ (<1)$ and $\widetilde{\mathsf{C}}_\beta = \sum_{n=0}^\infty \widetilde{\mathbf{P}}_\beta(n \in \tau)(<\infty)$.
(3) (The critical regime.) If we set $m_N := \sum_{n=0}^N \overline{K}(n)$ then $N \mapsto m_N$ is increasing and, if $\alpha \geq 1$, slowly varying and

$$Z_{N,\beta_c}^c \sim \Sigma_K \times \begin{cases} 1/m_N & \text{if } \alpha \geq 1, \\ \alpha\sin(\pi\alpha)N^{\alpha-1}/(\pi L(N)) & \text{if } \alpha \in (0,1), \end{cases} \qquad (2.15)$$

and for $K(\infty) > 0$

$$Z_{N,\beta_c}^f \sim \Sigma_K K(\infty) \times \begin{cases} N/m_N & \text{if } \alpha \geq 1, \\ \sin(\pi\alpha)N^\alpha/(\pi L(N)) & \text{if } \alpha \in (0,1), \end{cases} \qquad (2.16)$$

and the same holds for $\alpha = 0$ provided one replaces $\sin(\pi\alpha)/(\pi L(\cdot))$ with a suitable slowly varying function, cf. L.4, while for $K(\infty) = 0$ we have the identity

$$Z^{\mathrm{f}}_{N,\beta_c} = 1 \quad \text{for every } N. \tag{2.17}$$

Of course both in (2.15) and in (2.16) m_N may be replaced by m_∞ if $m_\infty < \infty$.

The key formula in the proof of Theorem 2.2 is:

$$Z^{\mathrm{c}}_{N,\beta} = \exp\left(\mathrm{F}(\beta)N\right)\widetilde{\mathbf{P}}_\beta\left(N \in \tau\right), \tag{2.18}$$

which is just a straightforward generalization of (1.9). The proof of Theorem 2.2 therefore boils down to estimating the renewal mass function of suitable renewal processes and notice that the three cases of the theorem correspond to (1) positive recurrent, (2) transient and (3) null recurrent, or (still) positive recurrent, renewal processes. Actually most of the work done here below is in estimating $Z^{\mathrm{f}}_{N,\beta}$, knowing the asymptotic behavior of $Z^{\mathrm{c}}_{N,\beta}$, which in turn follows directly from the results in Appendix A.5. In Appendix A.5 one can find full proofs except for the estimates needed to treat the critical case when they involve null recurrent renewals: for that we refer to the literature.

Proof of Theorem 2.2(1). Let us choose β such that $\mathrm{F}(\beta) > 0$. For the constrained case, the result is a direct consequence of the Renewal Theorem (Theorem A.3) and (2.18). For the free case we use the result of the constrained case to write

$$Z^{\mathrm{f}}_{N,\beta} = \sum_{n=0}^{N} Z^{\mathrm{c}}_{N-n,\beta}\overline{K}(n) + K(\infty)\sum_{n=0}^{N} Z^{\mathrm{c}}_{N-n,\beta} =: T_1 + T_2. \tag{2.19}$$

By the Dominated Convergence Theorem we have:

$$\exp(-N\mathrm{F}(\beta))T_2 = K(\infty)\sum_{n=0}^{N}\exp(-n\mathrm{F}(\beta))\widetilde{\mathbf{P}}_\beta\left(N-n \in \tau\right)$$

$$\underset{N\to\infty}{\sim} \frac{K(\infty)}{m_{\widetilde{K}_\beta}}\sum_{n=0}^{N}\exp(-n\mathrm{F}(\beta)) = \frac{K(\infty)}{m_{\widetilde{K}_\beta}\left(1 - \exp(-\mathrm{F}(\beta))\right)}, \tag{2.20}$$

where we have used the Renewal Theorem (Theorem A.3). Moreover (still by Theorem A.3)

$$T_1 = \exp\left(\mathrm{F}(\beta)N\right) \sum_{n=0}^{N} \exp\left(-\mathrm{F}(\beta)(N-n)\right) Z_{N-n,\beta}^{\mathrm{c}} \exp\left(-\mathrm{F}(\beta)n\right) \overline{K}(n)$$

$$\overset{N\to\infty}{\sim} \frac{1}{m_{\widetilde{K}_\beta}} \exp\left(\mathrm{F}(\beta)N\right) \sum_{n=0}^{\infty} \exp\left(-n\mathrm{F}(\beta)\right) \overline{K}(n)$$

$$= \frac{1}{m_{\widetilde{K}_\beta}} \exp\left(\mathrm{F}(\beta)N\right) \sum_{j=1}^{\infty} K(j) \frac{1 - \exp(-\mathrm{F}(\beta)j)}{1 - \exp(-\mathrm{F}(\beta))}$$

$$= \frac{1}{m_{\widetilde{K}_\beta}} \exp\left(\mathrm{F}(\beta)N\right) \frac{\Sigma_K - \exp(-\beta)}{1 - \exp(-\mathrm{F}(\beta))}.$$

$$(2.21)$$

Putting (2.20) and (2.21) together we get (2.12) with $a = \mathtt{f}$ and we are done.

Theorem 2.2(1)

$\square$

Proof of Theorem 2.2(2). Let us choose $\beta \in \overset{\circ}{\mathcal{D}}$. Observe that $\widetilde{K}_\beta(\cdot) = \varrho \widetilde{K}_{\beta_c}(\cdot)$. Recall (2.18) and apply Theorem A.4 to the (transient) renewal with inter-arrival law $\widetilde{K}_\beta(\cdot)$ to get

$$Z_{N,\beta}^{\mathrm{c}} = \widetilde{\mathbf{P}}_\beta\left(N \in \tau\right) \overset{N\to\infty}{\sim} \widetilde{K}_\beta(N) \frac{1}{(1-\varrho)^2}, \qquad (2.22)$$

which is the first statement in (2.14).

The second statement follows by arguing as follows. First we write

$$Z_{N,\beta}^{\mathtt{f}} = \sum_{n=0}^{N} Z_{n,\beta}^{\mathrm{c}} \overline{K}(N-n) + K(\infty) \sum_{n=0}^{N} Z_{n,\beta}^{\mathrm{c}} =: T_1 + T_2. \qquad (2.23)$$

By (2.18) we see that T_2 converges as $N \to \infty$ to $K(\infty)\widetilde{\mathtt{c}}_\beta$. Moreover

$$\frac{T_1}{\overline{K}(N)} = \sum_{n=0}^{N} Z_{n,\beta}^{\mathrm{c}} \frac{\overline{K}(N-n)}{\overline{K}(N)} = \left(\sum_{n=0}^{\lfloor N/2\rfloor} + \sum_{\lfloor N/2\rfloor+1}^{N}\right) Z_{n,\beta}^{\mathrm{c}} \frac{\overline{K}(N-n)}{\overline{K}(N)}$$

$$=: Q_N + R_N.$$

$$(2.24)$$

For Q_N we apply the Dominated Convergence Theorem by first observing that $\lim_{N\to\infty} \overline{K}(N-n)/\overline{K}(N) = 1$ for every n (this follows immediately for example from (A.18) or from (A.19)). Moreover, by the first of the two statements in (2.14), we have that $\sum_n Z^c_{n,\beta} < \infty$ and, on the other hand,

$$\sup_N \sup_{n \leq \lfloor N/2 \rfloor} \frac{\overline{K}(N-n)}{\overline{K}(N)} < \infty, \qquad (2.25)$$

which follows by applying the property L.3 (Appendix A.4), for $\alpha > 0$, and L.4 (Appendix A.4) for $\alpha = 0$, and by using the uniform convergence property of slowly varying functions (see (A.18)). Therefore by using (A.26), the fact that $\widetilde{K}_{\beta_c}(\cdot)$ is a probability density and the Fubini–Tonelli Theorem we obtain

$$\lim_{N \to \infty} Q_N = \sum_{n=0}^{\infty} Z^c_{n,\beta} = \sum_{n=0}^{\infty} \sum_{k=0}^{\infty} \varrho^k \widetilde{K}^{k*}_{\beta_c}(n) = \frac{1}{1-\varrho}. \qquad (2.26)$$

We are left with showing that $R_N = o(1)$. This can be done for $\alpha > 0$ by applying the rough bound L.1 of Appendix A.4 and the first of the two statements in (2.14):

$$R_N \leq c \sum_{n=\lfloor N/2 \rfloor + 1}^{N} K(n) \frac{\overline{K}(N-n)}{\overline{K}(N)} \leq c^2 N^{-\alpha+\varepsilon} \xrightarrow{N \to \infty} 0, \qquad (2.27)$$

for some $c > 1$ and $\varepsilon \in (0, \alpha)$. In the case $\alpha = 0$ we use L.4 of Appendix A.4 that tells us that $\overline{K}(\cdot)$ is slowly varying and therefore, by the uniform convergence property, $\overline{K}(N-n)/\overline{K}(N)$ is bounded uniformly in N and $n = \lfloor N/2 \rfloor + 1, \ldots, N$ and the proof is complete.

Theorem 2.2(2)
□

Proof of Theorem 2.2(3). In this case (2.18) says that Z^c_{N,β_c} is the renewal mass function of the recurrent renewal with inter-arrival discrete probability density $\widetilde{K}_{\beta_c}(\cdot) = K(\cdot)/\Sigma_K$. Of course the case $\sum_{n \in \mathbb{N}} nK(n) < \infty$, that is $m_{K_{\beta_c}} < \infty$ follows by a direct application of the Renewal Theorem (Theorem A.3) since $m_{\widetilde{K}_{\beta_c}} = m_\infty/\Sigma_K$.

In the case $\alpha = 1$ and $\sum_{n \in \mathbb{N}} nK(n) = \infty$ we still need the asymptotic behavior of the renewal mass function associated to the inter-arrival law $\widetilde{K}_{\beta_c}(\cdot)$, but the renewal theorem this time just tells us that Z^c_{N,β_c} vanishes for $N \to \infty$, since we are dealing with a null recurrent renewal. The sharper estimate claimed in (2.15) follows directly from Theorem A.6, equation (A.41).

If instead $\alpha \in (0,1)$ the asymptotic behavior of Z^c_{N,β_c} is given directly by Theorem A.7, applied once again to the null recurrent renewal with inter-arrival law $\widetilde{K}_{\beta_c}(\cdot)$. This completes the proof of (2.15).

In the free case we observe that, by the standard decomposition on the last renewal before (or equal to) N and (2.18), we have (with the convention that $\{j, \ldots, k\} = \emptyset$ if $j > k$)

$$
\begin{aligned}
Z^{\mathtt{f}}_{N,\beta_c} &= \sum_{n=0}^{N} Z^{\mathtt{c}}_{n,\beta_c} \mathbf{P}\left(\tau \cap \{1, \ldots, N-n\} = \emptyset\right) \\
&= \sum_{n=0}^{N} \widetilde{\mathbf{P}}_{\beta_c}\left(n \in \tau\right) \left(\overline{K}(N-n) + K(\infty)\right) \qquad (2.28) \\
&= \sum_{n=0}^{N} \widetilde{\mathbf{P}}_{\beta_c}\left(n \in \tau\right) \left(\Sigma_K \sum_{j > N-n} \widetilde{K}_{\beta_c}(j) + K(\infty)\right).
\end{aligned}
$$

Notice that in the recurrent case ($K(\infty) = 0$ and $\Sigma_K = 1$) $Z^{\mathtt{f}}_{N,\beta_c}$ coincides with $\mathbf{P}(\Omega_S)(= 1)$ and this proves (2.17). The same observation applied to the general case leads from (2.28) to

$$
Z^{\mathtt{f}}_{N,\beta_c} = \Sigma_K + K(\infty) \sum_{n=0}^{N} \widetilde{\mathbf{P}}_{\beta_c}\left(n \in \tau\right). \qquad (2.29)
$$

The integrated behavior of $\widetilde{\mathbf{P}}_{\beta_c}\left(n \in \tau\right)$ is given by Theorem A.6 and also (2.16) is therefore proven.

Theorem 2.2(3)
□

2.3 Path Properties: The Infinite Volume Limit

We now concentrate on the weak limit of the sequence $\left\{\mathbf{P}^{a}_{N,\beta}\right\}_N$. For the next statement we assume $\alpha > 0$ when dealing with $\mathbf{P}^{\mathtt{c}}_{N,\beta_c}$ and in all other cases there is no restriction on α (see Section 2.5).

Theorem 2.3 *For every β and for both $a = \mathtt{c}$ and $a = \mathtt{f}$ we have*

$$
\mathbf{P}^{a}_{N,\beta} \stackrel{N \to \infty}{\Longrightarrow} \widetilde{\mathbf{P}}_{\beta}. \qquad (2.30)
$$

The conciseness of this result is of course hiding a very rich phenomenology, that we sum up here:

- If $\beta > \beta_c$ the limiting renewal is positive recurrent.
- If $\beta < \beta_c$ the limiting renewal is transient (terminating).

- If $\beta = \beta_c$ the limiting renewal is recurrent, but it may be null recurrent (if $\sum_n nK(n) = \infty$) or positive recurrent (if $\sum_n nK(n) < \infty$).

Proof. For any $n \in \mathbb{N}$ and any $(t_1, \ldots, t_n) \in \mathbb{N}^n$ such that $t_1 < t_2 < \ldots < t_n \le N$ we write

$$\mathbf{P}^a_{N,\beta}(\tau_i = t_i \text{ for } i = 1, \ldots, n) =$$

$$\widetilde{\mathbf{P}}_\beta(\tau_i = t_i \text{ for } i = 1, \ldots, n) \left(\frac{\exp(\mathrm{F}(\beta)t_n)Z^a_{N-t_n,\beta}}{Z^a_{N,\beta}} \right), \quad (2.31)$$

and, by Theorem 2.2 we see that, in all regimes and for every $K(\cdot)$, the term between parentheses is of the form $R(N - t_n)/R(N)$, with $R(n) = n^\beta \widetilde{L}(n)$, for some $\beta \in \mathbb{R}$ and some slowly varying function $\widetilde{L}(\cdot)$. It is then clear that this term converges to 1 as $N \to \infty$ and the proof is complete. $\quad \square$

2.4 Path Properties: The Scaling Limit

Scaling limits deal with the large N behavior of the whole polymer. This is the reason for giving here the following result that deals with the *global* quantity *contact density*, that is $\mathrm{N}(\beta) = \mathrm{F}'(\beta)$, *cf.* (1.21), which is in general defined only for $\beta \ne \beta_c$.

Theorem 2.4 *For $\beta \ne \beta_c$ we have*

$$\mathrm{N}(\beta) = 1/m_{\widetilde{K}_\beta} \in [0,1). \tag{2.32}$$

Moreover

$$\mathrm{N}(\beta_c + \delta) \overset{\delta \searrow 0}{\sim} \begin{cases} \sum_{n \in \mathbb{N}} nK(n)/\Sigma_K & \text{if } \sum_{n \in \mathbb{N}} nK(n) < \infty, \\ \Sigma_K \delta^{(1-\alpha)/\alpha}\widetilde{L}(1/\delta)/\Gamma(1 - \alpha) & \text{if } \alpha \in [0,1), \end{cases}$$

$$(2.33)$$

where $\widetilde{L}(\cdot)$ is the slowly varying function defined by $\widetilde{L}(1/\delta) := \left(\hat{L}(1/\delta)\right)^{1-\alpha} / \left(L(1/\delta^{1/\alpha}\hat{L}(1/\delta))\right)$, with $\hat{L}(\cdot)$ given in Theorem 2.1 and the expression for $\alpha = 0$ has to be interpreted simply as saying that $\mathrm{N}(\beta_c + \delta)$ vanishes faster that any power of δ. If $\alpha = 1$ and $\sum_{n \in \mathbb{N}} nK(n) = \infty$ then $x \mapsto \mathrm{N}(\beta_c + (1/x))$ vanishes in a slowly varying fashion.

Proof. We start by observing that

$$\frac{\partial}{\partial \beta} \frac{1}{N} \log Z^{\mathrm{c}}_{N,\beta} = \mathbf{E}_{N,\beta} \left[\mathcal{N}_N(\tau)/N \right] \overset{N \to \infty}{\longrightarrow} \mathrm{N}(\beta), \qquad (2.34)$$

where the limit is guaranteed to exist, by convexity, if $\beta \neq \beta_c$, that is when $\mathrm{F}(\cdot)$ is differentiable (see Appendix A.1.1). Therefore $\mathrm{F}'(\beta) = \mathrm{N}(\beta)$ for $\beta \neq \beta_c$. Moreover

$$\mathbf{E}_{N,\beta} \left[\mathcal{N}_N(\tau)/N \right] = \frac{1}{\exp(-\mathrm{F}(\beta)N) Z^{\mathrm{c}}_{N,\beta}} \sum_{n=0}^{N} \sum_{\substack{\ell \in \mathbb{N}^n : \\ \sum_{i=1}^{n} \ell_i = N}} \frac{n}{N} \prod_{i=1}^{n} \tilde{K}_\beta(\ell_i).$$

$$(2.35)$$

In the denominator we recognize $\tilde{\mathbf{P}}_\beta(N \in \tau)$ and the expression in the numerator is equal to $\tilde{\mathbf{E}}_\beta \left[\mathcal{N}_N(\tau)/N; N \in \tau \right]$. If $\beta > \beta_c$ then $m_{\tilde{K}_\beta} < \infty$ and, by the Renewal Theorem, Theorem A.3, $\tilde{\mathbf{P}}_\beta(N \in \tau) \to 1/m_{\tilde{K}_\beta} > 0$ and by the law of large numbers $\mathcal{N}_N(\tau)/N \to 1/m_{\tilde{K}_\beta}$ $\tilde{\mathbf{P}}_\beta$–a.s.. Since of course $0 \leq \mathcal{N}_N(\tau)/N \leq 1$, by the Dominated Convergence Theorem we obtain (2.32). On the other hand, for $\beta < \beta_c$, $\mathrm{F}'(\beta) = 0$ and $m_{\tilde{K}_\beta} = \infty$. Therefore the proof of (2.32) is complete.

For what concerns (2.33), it follows directly (B.14) (and the lines that follow it for the case $\alpha = 1$) and from the asymptotic behavior of $\mathrm{F}(\cdot)$ close to β_c (Theorem 2.1). $\qquad\square$

Turning to scaling limits the results are in term of convergence in law of sequences of random sets, namely $\{(\tau/N) \cap [0,1]\}_N$, insisting on the fact that $(\tau/N) \cap [0,1] =: \tau_{(N)}$ is a (random) closed subset of $[0,1]$. The topology on the space of closed subsets is the Hausdorff (or Matheron) one, we refer to Appendix A.5.4 for precise definitions. The first result we present is the following

Theorem 2.5 *Assume that τ is distributed according to $\mathbf{P}^a_{N,\beta}$.*

(1) If $\beta > \beta_c$, that is $\beta \in \mathcal{L}$, then the sequence of random sets $\{\tau_{(N)}\}_N$ converges in law to the set $[0,1]$, both in the case $a = \mathtt{c}$ and $a = \mathtt{f}$.

(2) If $\beta < \beta_c$, that is $\beta \in \overset{\circ}{\mathcal{D}}$, then the sequence of random sets $\{\tau_{(N)}\}_N$ converges in law to the set $\{0\}$, in the case $a = \mathtt{f}$, and to $\{0,1\}$, in the case $a = \mathtt{c}$.

These are clear statements respectively of localization and of delocalization and they are proven in full generality. On the other hand the limit processes are trivial (and, due to that, the convergence is also in probability and not only in law).

Proof. If $\beta > \beta_c$, by Theorem 2.2(1) $\mathbf{P}_{N,\beta}^{\mathrm{f}}(N \in \tau) = Z_{N,\beta}^{\mathrm{c}}/Z_{N,\beta}^{\mathrm{f}}$ is bounded away from zero uniformly in N, so it is sufficient to prove the result for the free case. For $n \in \mathbb{N} \cup \{0\}$ and $n \notin \tau$ set $\mathrm{gap}_n(\tau) = \tau_{j(n)} - \tau_{j(n)-1}$, where $j(n) = \mathcal{N}_n(\tau) + 1$. Otherwise, *i.e.* if $n \in \tau$, set $\mathrm{gap}_n(\tau) = 0$. Define moreover $\mathrm{gap}_N^*(\tau) = \sup_{n<N} \mathrm{gap}_n(\tau)$. By the union bound we have for $c > 0$

$$\mathbf{P}_{N,\beta}^{\mathrm{c}}\left(\mathrm{gap}_N^*(\tau) > c \log N\right) \leq \sum_{n=0}^{N-1} \mathbf{P}_{N,\beta}^{\mathrm{c}}\left(\mathrm{gap}_n(\tau) > c \log N\right). \qquad (2.36)$$

Let us estimate the right-hand side by observing that for $0 \leq k < m \leq N$ we have

$$\mathbf{P}_{N,\beta}^{\mathrm{c}}\left(\tau \cap \{k,\ldots,m\} = \{k,m\}\right) = \frac{Z_{k,\beta}^{\mathrm{c}} K(m-k)\exp(\beta) Z_{N-m,\beta}^{\mathrm{c}}}{Z_{N,\beta}^{\mathrm{c}}}$$

$$\leq \frac{K(m-k)\exp(\beta)}{Z_{m-k,\beta}^{\mathrm{c}}} \overset{(m-k)\to\infty}{\asymp} \exp\left(-(m-k)\mathrm{F}(\beta)\right), \qquad (2.37)$$

where the inequality has been obtained by inserting in the denominator the event $\{\{k,m\} \in \tau\}$ and by applying the renewal property. Therefore we can find $c_1 > 0$ such that for every $M \in \mathbb{N}$ and every $n \leq N$

$$\mathbf{P}_{N,\beta}^{\mathrm{c}}\left(\mathrm{gap}_n(\tau) > M\right) \leq c_1 \exp(-M\mathrm{F}(\beta)/2). \qquad (2.38)$$

By plugging this estimate into (2.36) and by choosing $c = 3/\mathrm{F}(\beta)$ we get

$$\lim_{N\to\infty} \mathbf{P}_{N,\beta}^{\mathrm{c}}\left(\mathrm{gap}_N^*(\tau) > c \log N\right) = 0. \qquad (2.39)$$

So, with probability close to one, the maximal distance between the points in $\tau_{(N)}$ is $O(\log N/N)$ and Theorem 2.5(1) is proven.

If instead $\beta < \beta_c$ let us define $L_N(\tau) := \max(\tau \cap [0, N/2])$ and $R_N(\tau) := \min(\tau \cap [N/2, N])$ (we will be working in the constrained case, so $R_N(\tau)$ is well defined). For every $M_1, M_2 < N/2$ we have

$$\mathbf{P}_{N,\beta}^{\mathrm{c}}\left(L_N(\tau) \geq M_1,\, R_N(\tau) \leq N - M_2\right) =$$

$$\sum_{j,m\in E_M} \frac{Z_{j,\beta}^{\mathrm{c}} K(m-j)e^{\beta} Z_{N-m,\beta}^{\mathrm{c}}}{Z_{N,\beta}^{\mathrm{c}}}, \qquad (2.40)$$

where $\underline{M} := (M_1, M_2)$ and $E_{\underline{M}} := \{(j,m) : M_1 \leq j \leq N/2, \ N/2 \leq m \leq N - M_2\}$. By Theorem 2.2(2) we can find $c > 0$ such that the left-hand side in (2.40) is bounded above by

$$c \sum_{j,m:\in E_{\underline{M}}} \frac{K(j)K(m-j)K(N-m)}{K(N)} =: cB(N,\underline{M}). \qquad (2.41)$$

Let us write $B(N, \underline{M})$ as $B_1(N, \underline{M}) + B_2(N, \underline{M})$ in which $B_1(N, \underline{M})$ is defined by restricting the sum in (2.41) to $j < N/4$ and $k > 3N/4$, and $B_2(N, \underline{M})$ is the rest. Let us estimate $B_2(N, \underline{M})$: if $j \geq N/4$ then $K(j)/K(N)$ is bounded above by a positive constant, uniformly in N (we are using here the uniform convergence property (A.18) of slowly varying functions). We are therefore left with estimating

$$\sum_{j,m:\in E_{\underline{M}}} K(m-j)K(N-m) \leq \sum_{\substack{j: \\ N/4 \leq j \leq N/2}} (K*K)(N-j) \leq cN(K*K)(N),$$

$$(2.42)$$

which vanishes as N tends to infinity (for the case $\alpha = 0$ observe that, by the properties of slowly varying functions $N(K*K)(N) \leq const.L(N)$ and one may show that in this case $L(\cdot)$ vanishes at infinity either directly or by applying L.4 of Appendix A.4). The same estimate holds if we assume $m \leq 3N/4$. For $B_1(N, \underline{M})$ we use the fact that $K(m-j)/K(N) \leq C$ uniformly in N, so

$$B_1(N, \underline{M}) \leq C \prod_{i=1,2} \Big(\sum_{j: \ M_i \leq j < N/4} K(j) \Big) \leq C\overline{K}(M_1-1)\overline{K}(M_2-1). \quad (2.43)$$

Therefore overall we have that

$$\limsup_{N\to\infty} \mathbf{P}^{c}_{N,\beta}\left(L_N(\tau) \geq M_1, \ R_N(\tau) \leq N - M_2\right) \leq$$

$$const.\overline{K}(M_1)\overline{K}(M_2) \overset{|\underline{M}|\to\infty}{\longrightarrow} 0. \quad (2.44)$$

This implies immediately that the limit of $\{\tau_{(N)}\}_N$ is $\{0,1\}$.

In the free case one uses the same idea: the net result is

$$\limsup_{N\to\infty} \mathbf{P}^{c}_{N,\beta}\left(L_N(\tau) \geq M\right) \leq const. \ \overline{K}(M), \qquad (2.45)$$

with the new definition $L_N(\tau) := \max \tau \cap [0, N]$, both in the case in which τ is transient and in the case in which it is recurrent, and the conclusion follows. $\square$

Remark 2.6 *It is clear that (2.39), (2.44) and (2.45) are quantitative estimates that go much beyond what is stated in Theorem 2.5.*

In the critical case the scaling limits are much richer and we are going to discuss them in full generality (see Rem. 2.8 and Section 2.5), but the results that we are going to present in detail are more limited and they should just be taken as a sample of what one can obtain:

Theorem 2.7 *If τ is distributed according to $\mathbf{P}^{\mathrm{f}}_{N,\beta_c}$ then*

(1) if $K(\infty) = 0$ then $\{\tau_{(N)}\}_N$ converges in law toward $\mathcal{A}_{\min(\alpha,1)} \cap [0,1]$ where $\mathcal{A}_\gamma$ is the regenerative set process with Lévy exponent $\gamma \in [0,1]$;
(2) if $K(\infty) > 0$ and $\alpha \in (0,1)$ then $\{\tau_{(N)}\}_N$ converges in law toward $\widetilde{\mathcal{A}}_\alpha \subset [0,1]$, where for $\alpha \in (0,1)$ the law of $\widetilde{\mathcal{A}}_\alpha$ is absolutely continuous with respect to the law of $\mathcal{A}_\alpha \cap [0,1]$ with Radon–Nykodym density equal to $(\alpha\pi/\sin(\alpha\pi))(1 - \max(\mathcal{A}_\alpha \cap [0,1]))^\alpha$. In the case $K(\infty) > 0$ and $\sum_{n \in \mathbb{N}} nK(n) < \infty$ then $\{\tau_{(N)}\}_N$ converges in law toward toward $[0,U]$, where U is uniformly distributed in $[0,1]$.

We stress that $\mathcal{A}_0 = \{0\}$ and that $\mathcal{A}_1 = [0,\infty)$, but $\mathcal{A}_\alpha$ for $\alpha \in (0,1)$ is a highly non-trivial random set which coincides, for example, with the zero level set of a Brownian motion if $\alpha = 1/2$: for definitions and properties of regenerative set processes see Appendix A.5.4. We stress also that the density term that is present when $\alpha \in (0,1)$ and the underlying renewal is transient ($\Sigma_K < 1$) is precisely a term that pushes the rightmost point of the limit process away from 1 and closer to the origin: it is clearly a reminiscence of the transient nature of the original process. This repulsion is even more evident if $\sum_n nK(n) < \infty$.

Remark 2.8 *In the constrained endpoint case a part of the arguments is actually particularly limpid. In fact one readily realizes that for $n \in \mathbb{N}$ and $\underline{t} \in \mathbb{N}^n$, with $(t_0 :=)0 < t_1 < \ldots < t_{n-1} \leq N =: t_n$, one has the expression*

$$\mathbf{P}^{\mathrm{c}}_{N,\beta_c}\left(\tau \cap [1,N] = \underline{t}\right) = \frac{\prod_{j=1}^n \widetilde{K}_{\beta_c}(t_j - t_{j-1})}{\widetilde{\mathbf{P}}_{\beta_c}(N \in \tau)}. \tag{2.46}$$

In a more compact fashion this is saying nothing but

$$\mathbf{P}^{\mathrm{c}}_{N,\beta_c}\left(\tau \cap [1,N] = A\right) = \widetilde{\mathbf{P}}_{\beta_c}\left(\tau \cap [1,N] = A \mid N \in \tau\right), \tag{2.47}$$

for every $A \subset \{1, \ldots, N\}$. Hence the critical scaling limit in the constrained case is just the generalization of Proposition A.8 to the case in which the renewal is constrained to passing through N (see Section 2.5).

Proof of Theorem 2.7. If $K(\infty) = 0$ then $\beta_c = 0$ and $\mathbf{P}^f_{N,\beta_c}$ is $\mathbf{P}$ (for every N). Point (1) is therefore just Theorem A.8.

If $K(\infty) > 0$ and $\alpha \in (0,1)$ we argue as follows: for $\underline{t} \in \mathbb{N}^n$, $(t_0 :=)0 < t_1 < \ldots < t_n \leq N$ we write

$$\mathbf{P}^f_{N,\beta_c}(\tau_1 = t_1, \ldots, \tau_n = t_n, \tau_{n+1} > N) = \frac{(\overline{K}(N - t_n) + K(\infty))Z^c_{N-t_n,\beta_c}}{Z^f_{N,\beta_c}}$$

$$= \widetilde{\mathbf{P}}_{\beta_c}(\tau_1 = t_1, \ldots, \tau_n = t_n, \tau_{n+1} > N) \frac{\overline{K}(N - t_n) + K(\infty)}{Z^f_{N,\beta_c} \sum_{j>N-t_n} \widetilde{K}_{\beta_c}(j)}. \quad (2.48)$$

If F is a bounded measurable function on $\mathcal{C}_1$ (the set of closed subsets of $[0,1]$ equipped with the Hausdorff metric) then we have

$$\mathbf{E}^f_{N,\beta_c}[F(\tau_{(N)})] = \widetilde{\mathbf{E}}_{\beta_c}\left[F(\tau_{(N)}) \frac{\overline{K}(N(1 - \max(\tau_{(N)}))) + K(\infty)}{Z^f_{N,\beta_c} \sum_{j>N(1-\max(\tau_{(N)}))} \widetilde{K}_{\beta_c}(j)}\right].$$
$$(2.49)$$

Let us take the limit in this expression by first inserting the event $\max \tau_{(N)} < (1 - \varepsilon)$, $\varepsilon > 0$. On this event, by Theorem 2.2 and then by the uniform convergence property of slowly varying functions, we have

$$\frac{\overline{K}(N(1 - \max(\tau_{(N)}))) + K(\infty)}{Z^f_{N,\beta_c} \sum_{j>N(1-\max(\tau_{(N)}))} \widetilde{K}_{\beta_c}(j)} \overset{N\to\infty}{\sim} \frac{\alpha\pi}{\sin(\alpha\pi)}(1 - \max(\tau_{(N)}))^\alpha$$
$$(2.50)$$

uniformly in the trajectories of τ. Theorem A.8 says that the law of $\tau_{(N)}$ (τ is distributed according to $\widetilde{\mathbf{P}}_{\beta_c}$) converges to the law of $\mathcal{A}_\alpha$, which implies the convergence of the law of $\max(\tau_{(N)})$ to the law of $\max \mathcal{A}_\alpha$. Since $\max \mathcal{A}_\alpha < 1$ almost surely, by taking $\varepsilon \searrow 0$ we complete the proof in the case $\alpha \in (0,1)$.

In the case $\sum_n nK(n) < \infty$ we start by computing the distribution of $\max(\tau_{(N)})$: by (2.49), by Theorem 2.2(3) and by decomposing the expectation in the right-hand side of (2.49) on the events $\{N \max(\tau_{(N)}) = n\}$ we have for $t < 1$

$$\mathbf{P}^f_{N,\beta_c}(\max(\tau_{(N)}) < t) \overset{N\to\infty}{\sim} \frac{m_N/\Sigma_K}{N} \sum_{n<tN} \widetilde{\mathbf{P}}_{\beta_c}(n \in \tau). \quad (2.51)$$

By Theorem A.6 and by the fact that $N \mapsto m_N$ is slowly varying we have $\sum_{n<tN} \widetilde{\mathbf{P}}_{\beta_c}(n \in \tau) \sim Nt/(m_N/\Sigma_K)$. Therefore

$$\lim_{N \to \infty} \mathbf{P}^{\mathrm{f}}_{N,\beta_c}\left(\max(\tau_{(N)}) < t\right) = t. \tag{2.52}$$

To complete the proof it is sufficient to show that for every $\varepsilon > 0$

$$\lim_{N \to \infty} \mathbf{P}^{\mathrm{f}}_{N,\beta_c}\left(\mathrm{gap}^*_{\max(N\tau_{(N)})}(\tau) > \varepsilon N\right) = 0. \tag{2.53}$$

By the formula

$$\mathbf{P}^{\mathrm{f}}_{N,\beta_c}\left(\mathrm{gap}^*_{\max(N\tau_{(N)})}(\tau) > \varepsilon N\right) =$$
$$\sum_n \mathbf{P}^{\mathrm{c}}_{n,\beta_c}\left(\mathrm{gap}^*_n(\tau) > \varepsilon N\right) \mathbf{P}^{\mathrm{f}}_{N,\beta_c}\left(\max(N\tau_{(N)}) = n\right). \tag{2.54}$$

It suffices to show that $\lim_{N \to \infty} \sup_{n \in (\varepsilon N, N]} \mathbf{P}^{\mathrm{c}}_{n,\beta_c}\left(\mathrm{gap}^*_n(\tau) > \varepsilon N\right) = 0$ and, by recalling (2.47), if $m_{\widetilde{K}_{\beta_c}} < \infty$ this bound is implied by

$$\sup_{n \in (\varepsilon N, N]} \widetilde{\mathbf{P}}_{\beta_c}\left(\mathrm{gap}^*_n(\tau) > \varepsilon N\right) \leq N\widetilde{\mathbf{P}}_{\beta_c}\left(\tau_1 > \varepsilon N\right), \tag{2.55}$$

and the fact that the last term tends to 0 as $N \to \infty$ (this is immediate if $\alpha > 1$ and it follows from L.4 of Appendix A.4 if $\alpha = 1$).

Theorem 2.7 □

2.5 Bibliographic Complements

Complements on Section 2.1

This is a sharpening, obtained by renewal theory arguments, of the classical computation performed for example in [Fisher (1984), Section 6]. It is a sharpening because it includes the slowly varying corrections to polynomial scaling.

Complements on Section 2.2

Sharp estimates on the partition function of very general random walk based models ($\alpha = 1/2$) can be found in [Caravenna *et al.* (2006a)]. For (p,q)-walks we mention [Isozaki and Yoshida (2001)]. In this section we generalize the approach in [Caravenna *et al.* (2006a)] to general α and we heavily rely on renewal mass function estimates, treated in detail Appendix A.5. Of particular importance are the sharp results in [Doney (1997)] that allow an

almost complete treatment of the constrained case. We are saying *almost*
because probably to the attentive reader it has not escaped that the result
in Theorem 2.2(3), for the constrained case, does not include $\alpha = 0$. This is
a result missing in the theory of local estimates for mass renewal functions:
it is unclear whether or not a serious effort has been put into obtaining
such a result [Doney, Private Communication (2006)].

Complements on Section 2.3

This section generalizes the work in [Caravenna *et al.* (2006a)], which in
turn generalized for example [Isozaki and Yoshida (2001)] and [Deuschel
et al. (2005)]. Exact computations in this direction can also be found in
the physical literature, see [Burkhardt (1981)] and [van Leeuwen, J.M.J.,
Hilhorst (1981)]. We point out also [Roynette *et al.* (2003)], that focuses instead on exponential perturbations of Brownian motion. Still in
the Brownian context, [Funaki (2005)] derives scaling limits for a case in
which boundary conditions, chosen to favor paths in the upper half-plane,
are in competition with a bulk potential that favors the lower half-plane. Of
course between what we are doing and the papers we have just mentioned
there is the fundamental difference that we are dealing only with renewal
processes, and not with the returns to zero of a random walk, or another
process. We stress however that for the cases we have mentioned it is not
so difficult to go from a weak convergence result of the zero level set to the
convergence of the full process. This is due to the fact that, conditionally
on the location of the zeros, the trajectory of the process in between is just
an unperturbed random walk excursion.

   Let us point out an interesting open question on the infinite volume
limit measure $\widetilde{\mathbf{P}}_\beta$ for $\beta > \beta_c$. The two point correlation function is

$$c_\beta(n,j) := \widetilde{\mathbf{P}}_\beta\left(\{n,j\} \subset \tau\right) - \widetilde{\mathbf{P}}_\beta\left(n \in \tau\right)\widetilde{\mathbf{P}}_\beta\left(j \in \tau\right). \qquad (2.56)$$

We ask what the asymptotic behavior of $c_\beta(n, n + k)$ is for k large and
$\beta > \beta_c$, at least in the Laplace sense. It is well known for example that
for every $\beta > \beta_c$ there exists $\varepsilon > 0$ such that $|c_\beta(n, n + k)| \exp(-\varepsilon k)$, for k
sufficiently large. This follows from the classical result on the convergence
to equilibrium of renewal processes (see [Ney (1981)] but also [Asmussen
(2003)]): in fact, using, as usual, δ_n for $\mathbf{1}_{n \in \tau}$, $c_\beta(n, n + k)$ is equal to
$\widetilde{\mathbf{E}}_\beta[\delta_n]$ times $\widetilde{\mathbf{E}}_\beta[\delta_{n+k}|\delta_n = 1] - \widetilde{\mathbf{E}}_\beta[\delta_{n+k}]$ and it is therefore this last term
that determines the speed of decay of correlations. Sharp answers to this
question are still missing, in spite of the substantial effort that has been

put into it, we cite in particular [Lund and Tweedy (1996)] and [Berenhaut and Lund (2002)]. For the inter-arrivals on which we concentrate, general results do not seem to exist. In particular cases however precise estimates can be made: for example if τ is the zero level set of a simple random walk one can show that $c_\beta(n, n+k) \asymp \exp(-k\mathrm{F}(\beta))$, as $k \to \infty$. This follows from a result in [Lund and Tweedy (1996)], but also from [Toninelli (2006)] (see also Section 7.4): in any case the proof depends on a particular stochastic ordering property. But does such an appealing result hold for more general processes? This is not clear at all and, possibly, it is more reasonable to expect a result like $c_\beta(n, n + k) \overset{k\to\infty}{\asymp} \exp(-kf(\beta))$, with $f(\beta) \sim \mathrm{F}(\beta)$ as $\beta \searrow \beta_c$ or even only $\log f(\beta) \sim \log \mathrm{F}(\beta)$.

Complements on Section 2.4

Again, this section generalizes [Isozaki and Yoshida (2001)], [Deuschel *et al.* (2005)] and [Caravenna *et al.* (2006a)], see also [Upton (1999)] and references therein for non-rigorous results on scaling limits. Once again, we focus only on the zero level set and our result is given in terms of convergence in the Hausdorff topology. Even if the results are non-trivial in all regimes, from the point of view of the limiting processes, they are non-trivial only at criticality where the regenerative set processes with Lévy exponent $\gamma \in (0,1)$ come up. As already stressed in the text, the results are only partial, but they clearly show the relevance of the stable regenerative set processes in describing the criticality of the general class of models we consider: critical behavior that is characterized by the exponent α alone. In [Deuschel *et al.* (2005)] and [Caravenna *et al.* (2006a)] the results are restricted to the case $\alpha = 1/2$ and to the wetting case, but the analysis of the full trajectories of the polymer (or interface) is worked out and the limit process arising at criticality is the absolute value of the Brownian motion (called reflected Brownian motion), respectively Brownian bridge, in the free endpoint case, respectively in the constrained endpoint case. Outside of criticality instead the scaling is non-trivial only in the delocalized phase and one observes the so called Brownian meander process, for the free endpoint case, and the Brownian bridge conditioned to stay positive, for the constrained endpoint case (See [Revuz and Yor (1999)] for precise definitions of these processes). This result had been already proven for (p, q)-walks in [Isozaki and Yoshida (2001)], and it had been conjectured at several instances in the physical literature (see [Upton (1999)] and references therein). However we point out that for (p, q)-walks the reflected

Brownian motion is *essentially* already present at *microscopic level*, that is at the level of random walks: it is in fact rather easy to see, for example, that at criticality the wetting model based on the simple random walk is precisely the reflected simple random walk even at finite volume. This precise correspondence disappears beyond (p, q)-walks and the techniques of proof employed are substantially different and they are in fact based on two steps:

(1) Establishing the scaling limit of the zero level set. Of course in the free case the scaling limit is $A_{1/2} \cap [0, 1]$. In the constrained case it is instead the random set process $A_{1/2} \cap [0, 1]$ *constrained to contain the point* 1. This is an imprecise definition: more precisely one shows convergence toward $(A_{1/2} \cap [0, 1]) / \max(A_{1/2} \cap [0, 1])$, a random set which of course contains 1. The proof however is technically more involved than for the free case (this comment is naturally linked to Rem. 2.8).

(2) A second step is *pasting* the random walk excursions: the idea is simply that given $t \in (0, 1]$, the probability that $t \in A_{1/2} \cap [0, 1]$ is zero. At a microscopic level this means that with large probability $\lfloor tN \rfloor$ is far, order of N, from a zero of the process. So the random walk excursion can be approximated by a Brownian excursion and one ends up with the process built by attaching (positive) Brownian excursions to the zero level set of a Brownian motion, that is the reflected Brownian motion. Brownian limits of random walks with positivity constraint is a classical research domain which is still actively evolving, see [Bryn-Jones and Doney (2004)], [Caravenna (2005b)] and references therein.

So what we are doing in Section 2.4 is step (1), but only for the free case. The extension to the constrained case is work in progress, as well as the extension of (2): in this case one has of course to analyze the scaling limit of models based on random walks with increments in the domain of attraction of stable laws.

Another important remark is that all the works we have mentioned are based on what we may call *sharp renewals*: either because one deals with (p, q)-walks, or because one deals with discrete walks in the presence of a hard wall, or because one deals with more general increments but the pinning is the so called δ-pinning (see *e.g.* [Deuschel *et al.* (2005)]), the walk crosses the interface by touching it and this automatically leads, at a microscopic level, to a renewal structure. Such a structure is not present for example in the following very natural wetting model:

$$\frac{d\mathbf{P}_{N,\beta,a}}{d\mathbf{P}}(S) := \frac{1}{Z_{N,\beta,a}} \exp\left(\beta \sum_{n=1}^{N} \mathbf{1}_{S_n \in [0,a]}\right) \mathbf{1}_{\{S_n \geq 0,\, n=1,2,...,N\}}, \quad (2.57)$$

where $a > 0$ and S, under $\mathbf{P}$, is a symmetric random walk with continuous increments in the normal domain of attraction (say: $\mathbf{E}[S_1^2] = 1$ and density of S_1 positive in a neighborhood of 0). It is not difficult to see that also such a model has a localization transition and that for every $a > 0$ there exists a critical point $\beta_c > 0$. It is natural to conjecture that, no matter what the value of a is, the scaling limits are the same as the ones found in [Caravenna *et al.* (2006a)]. The techniques we have discussed do not apply since the sequence of successive entry times in $[0, a]$ is not a renewal sequence. One would have probably to deal with the sequence of entry times and the location of the walk in $[0, a]$ at the entry moment (what is usually called the *overshoot*). This, however, is an open problem.

Chapter 3

Weakly Inhomogeneous Models

In this chapter we consider polymer models with periodic charges. Precisely we consider the *copolymer with adsorption* model of Section 1.6.3, that is the measure $\mathbf{P}^a_{N,\omega,\underline{v}}$ defined by (1.66). This of course includes polymer and pinning models as particular cases. Clearly they are inhomogeneous models, but only *weakly* since they are homogeneous when looked upon in steps of length $T(\omega)$, *cf.* Definition 1.9. Since the degree of inhomogeneity grows with $T(\omega)$ these models could be of help in order to get a deeper understanding of fully inhomogeneous models, see in particular Section 4.3 and the bibliographic complements (Section 3.5).

In this chapter the aim is above all to stress the basic steps that allow to reduce the weakly inhomogeneous case to the homogeneous one and convey the idea that in this class of models one gets as far as in the homogeneous pinning model treated in Chapter 2. This *reduction* is not without a price, since in reality formulas become substantially more complex in the details. However it should be stressed from now that new phenomena appear in this set up and one of our purposes is to point out in an informal way the phenomenological richness of periodic models.

For the sake of conciseness we will work only with (p,q)-walks so $K(n) \overset{n\to\infty}{\sim} c_K n^{-3/2}$, *i.e.* $\alpha = 1/2$ (see Appendix A.6).

3.1 A Formula for the Free Energy: Reduction to a Finite Dimension Problem

In this section we will make some algebraic manipulations on the partition function of periodic models that lead to a formula for the free energy. This is the generalization of Proposition 1.1 and of formula (1.6) to weakly inhomogeneous models. Very much like in that case, the computation of the free energy is just a warm-up since the basic idea developed in this section naturally leads to sharp estimates.

In order to make formulas more compact we write the model as

$$\frac{d\mathbf{P}^a_{N,\omega}}{d\mathbf{P}}(S) = \frac{\mathbf{1}_{\Omega^a_S(N)}}{Z^a_{N,\omega}} \exp\left(\sum_{n=1}^{N} \hat{\omega}_n \Delta_n + \sum_{n=1}^{N} \omega_n \delta_n\right), \qquad (3.1)$$

which of course corresponds simply to the change of variables $-2\lambda(\hat{\omega}_n + h) \rightarrow \hat{\omega}_n$ and $\left(\beta\omega_n - \widetilde{h}\right) \rightarrow \omega_n$. With this choice $0 \geq \sum_{n=1}^{T} \hat{\omega}_n/T =:$ H and $T := T(\omega)$ (we assume $T > 1$). As it has already been shown in Section 1.6, establishing the existence of the free energy for periodic models is just a matter of exploiting the super-additive property of $\{\log Z^c_{nT,\omega}\}_n$:

$$\mathrm{F}_\omega := \lim_{N\to\infty} \frac{1}{N} \log Z^c_{N,\omega}, \qquad (3.2)$$

and, as usual, $\mathrm{F}_\omega \geq 0$ and the localized, respectively delocalized, regime is characterized by $\mathrm{F}_\omega > 0$, respectively $\mathrm{F}_\omega = 0$. The representation formula that we are deriving in this section yields in particular an alternative proof of the existence of the limit in (3.2).

For $n_1 < n_2$ and $n_1 \geq 0$ we introduce

$$\Psi^\omega(n_1, n_2) := \begin{cases} \omega_{n_2} + \hat{\omega}_{n_2} & \text{if } n_2 = n_1 + 1, \\ \omega_{n_2} + \varphi\left(\omega(n_1, n_2]\right) & \text{if } n_2 \geq n_1 + 2, \\ 0 & \text{otherwise}, \end{cases} \qquad (3.3)$$

where $\varphi(t) = \log((1 + \exp(-t))/2)$. By conditioning on τ and integrating over the sign of the excursions $\{s_i\}_i$, like in (1.64), we get

$$Z^c_{N,\omega} = \mathbf{E}\left[\exp\left(\sum_{j=1}^{\mathcal{N}_N(\tau)} \Psi^\omega(\tau_{j-1}, \tau_j)\right); \tau_{\mathcal{N}_N(\tau)} = N\right]. \qquad (3.4)$$

We have chosen $\Delta_n = 1$ when $S_{n-1} = S_n = 0$, but generalizations are of course straightforward.

The key point now is the following manipulation on $\Psi^\omega(n_1, n_2)$ that uses the periodic character of ω. We observe that

$$\hat{\omega}(n_1, n_2] = \sum_{n=n_1+1}^{n_2} \hat{\omega}_n = \Sigma_{[n_1],[n_2]} + (n_2 - n_1)\mathrm{H}, \qquad (3.5)$$

where Σ is a matrix on $\mathbb{S} := \mathbb{Z}/T\mathbb{Z}$ (an Abelian group with operation $+$), that is the set $\{1, \ldots, T\}$ with periodic boundary. With some abuse of notation we will often interpret $\alpha \in \mathbb{S}$ as an element of $\{1, \ldots, T\}$ (or $\mathbb{Z}$): in this sense if we write n as $jT + m$, $j \in \mathbb{N} \cup \{0\}$ and $m \in \{1, \ldots, T\}$ then $[n] = m$. So if we introduce for $\alpha, \beta \in \mathbb{S}$ and $n \in \mathbb{N}$

$$\Phi_{\alpha,\beta}^{\omega}(n) = \begin{cases} \omega_\beta + \hat{\omega}_\beta & \text{if } n = 1 \text{ and } n \in \beta - \alpha, \\ \omega_\beta + \varphi\left(\Sigma_{\alpha,\beta} + n\mathrm{H}\right) & \text{if } n \geq 2 \text{ and } n \in \beta - \alpha, \\ 0 & \text{otherwise,} \end{cases} \quad (3.6)$$

then $\Psi^\omega(n_1, n_2) = \Phi_{[n_1],[n_2]}^{\omega}(n_2 - n_1)$.

We define also

$$M_{\alpha,\beta}^{\omega}(n) := K(n)\mathbf{1}_{[n]=\beta-\alpha} \exp\left(\Phi_{\alpha,\beta}^{\omega}(n)\right). \quad (3.7)$$

The dependence on ω will often be dropped and we use the notation $(M^{2*})_{\alpha,\beta}(n) = (M * M)_{\alpha,\beta}(n) := \sum_{m,\gamma} M_{\alpha,\gamma}(m) M_{\gamma,\beta}(n - m)$. Notice that this sum is empty if $n = 1$ and for $n \geq 2$ the sum ranges from $m = 1$ to $m = n - 1$. Moreover $(M^{k*})_{\alpha,\beta}(n) = 0$ for $k > n$.

Lemma 3.1 (**Matrix representation**). *We have*

$$Z_{N,\omega}^{\mathrm{c}} = \sum_{k \in \mathbb{N}} \left(M^{k*}\right)_{[0],[N]}(N). \quad (3.8)$$

Proof. It is just the generalization of the basic formula (1.8). Observe in fact that

$$Z_{N,\omega}^{\mathrm{c}} = \sum_{k=1}^{N} \sum_{\substack{n_1,\ldots,n_k \in \mathbb{N}: \\ 0=:n_0<n_1<\ldots<n_k=N}} \prod_{j=1}^{k} K(n_j - n_{j-1}) \exp\left(\Phi_{[n_{j-1}],[n_j]}^{\omega}(n_j - n_{j-1})\right),$$

$$(3.9)$$

and, with our matrix notations, (3.8) follows. $\square$

Inspired by the procedure applied for homogeneous polymers we introduce an exponential modification also in this case: for $b \geq 0$ we introduce the matrix

$$A_{\alpha,\beta}^{b}(n) := M_{\alpha,\beta}(n) \exp(-bn). \quad (3.10)$$

We denote by $\lambda_\star^\omega(b)(= \lambda_\star(b))$ the Perron–Frobenius eigenvalue of the matrix $\sum_n A^b(n)$ which has positive entries (and it is therefore irreducible). By the general Perron–Frobenius theory (*cf.* Appendix A.8), the map $\lambda_\star^\omega :$ $[0,\infty) \to (0,\infty)$ is smooth and increasing, in particular $\lambda_\star^\omega(b) \geq \lambda_\star^\omega(0)$.

Theorem 3.2 *We have the formula*

$$F_\omega = \begin{cases} (\lambda_\star^\omega)^{-1}(1) & \text{if } \lambda_\star^\omega(0) > 1, \\ 0 & \text{if } \lambda_\star^\omega(0) \leq 1. \end{cases} \tag{3.11}$$

Proof. We are going to treat the cases of $\lambda_\star^\omega(0) > 1$, $\lambda_\star^\omega(0) = 1$ and $\lambda_\star^\omega(0) < 1$ separately.

The case of $\lambda_\star^\omega(0) > 1$.

In this case there exists a unique solution $b > 0$ to $\lambda_\star^\omega(b) = 1$. We want to show that $b = F_\omega$. Let us call ξ the right eigenvector of the matrix $\sum_n A^b(n)$. The normalization of ξ may be chosen at like. Set

$$\Gamma_{\alpha,\beta}(n) := \exp(-bn)M_{\alpha,\beta}(n)\frac{\xi_\beta}{\xi_\alpha}, \tag{3.12}$$

and note that $\sum_{\beta,n} \Gamma_{\alpha,\beta}(n) = 1$ along with

$$
\begin{aligned}
(\Gamma * \Gamma)_{\alpha,\beta}(n) &= \exp(-bn) \sum_{m,\gamma} M_{\alpha,\gamma}(m)\frac{\xi_\gamma}{\xi_\alpha} M_{\gamma,\beta}(n-m)\frac{\xi_\beta}{\xi_\gamma} \\
&= \exp(-bn)\frac{\xi_\beta}{\xi_\alpha}(M * M)_{\alpha,\beta}(n).
\end{aligned}
\tag{3.13}
$$

It is now easy to rewrite the formula in Lemma 3.1 in terms of Γ:

$$Z_{N,\omega}^c = \exp(bN)\frac{\xi_{[0]}}{\xi_{[N]}} \sum_{k \in \mathbb{N}} (\Gamma^{k*})_{[0],[N]}(N). \tag{3.14}$$

This formula has a probabilistic interpretation via *Markov renewal processes*. Namely, consider the process $(J,\tau) := \{(J_k,\tau_k)\}_{k=0,1,\dots}$ on $\mathbb{S} \times (\{0\} \cup \mathbb{N})$ built in two steps:

(1) First sample J, starting from $J_0 = 0$ (but one can of course generalize to $J_0 = \alpha \in \mathbb{S}$), by using the $\mathbb{S} \times \mathbb{S}$ irreducible stochastic matrix $\sum_{n \in \mathbb{N}} \Gamma(n)$.

(2) Then sample $\eta = \{\eta_j\}_{j\in\mathbb{N}}$, independent random variables with (non-identical) distribution: the probability that η_k is equal to n given $(J_{k-1}, J_k) = (\alpha, \beta)$ is $\Gamma_{\alpha,\beta}(n)/\sum_n \Gamma_{\alpha,\beta}(n)$. Then τ is just the partial sum process associated to η, *i.e.* $\tau_0 = 0$ and $\eta_j = \tau_j - \tau_{j-1}$. If $J_0 = \alpha$ it is natural to choose $\tau_0 \in \alpha$ (for definiteness, choose τ_0 in $\{0, 1, \ldots, T-1\}$).

We call τ, but also (J, τ), Markov renewal process [Asmussen (2003)]. It is a natural generalization of standard renewals, and we denote the law of (J, τ) by $\mathbf{P}_\alpha^\omega$, when $J_0 = \alpha$ (note that τ_0 is uniquely defined in all cases). It is immediate to see that (J, η) is a Markov chain with transition probabilities

$$\mathbf{P}_\gamma^\omega\big((J_{k+1}, \eta_{k+1}) = (\beta, n)\big|\ (J_k, \eta_k) = (\alpha, m)\big)\ =\ \Gamma_{\alpha,\beta}(n), \qquad (3.15)$$

for $k \in \mathbb{N} \cup \{0\}$, but if $k = 0$ we require $\alpha = \gamma$ and η_0 is arbitrarily defined since it does not affect the sampling of (J_1, η_1) and the definition of τ. We call $\Gamma(\cdot)$ the *kernel* of the Markov renewal.

Remark 3.3   *While τ is not a standard renewal, it is useful to observe that for $\alpha \in \mathbb{S}$ the point process $\tau^\alpha := \{\tau_j : j \in \mathbb{N}, J_j = \alpha\}$ is a standard renewal (delayed, if $\alpha \neq [0]$: add the point 0 to the sequence if the delay is 0 to match our definition of non-delayed renewal) and $\tau^\alpha - \tau_1^\alpha = \{\tau_j^\alpha - \tau_1^\alpha\}_j$ is a standard renewal. Notice however that $\tau^\alpha - \tau_1^\alpha \subset T(\mathbb{N} \cup \{0\})$ and $(\tau^\alpha - \tau_1^\alpha)/T$ is again a renewal with the property of being aperiodic. In general it is not easy to compute explicitly the inter-arrival distribution of τ^α. However the inter-arrival law of the process of returns to α of the finite state space Markov chain J is stochastically bounded by a geometric variable and, since there are only finitely many different distributions of inter-arrivals for the Markov renewal τ and since they are all integrable (they are even exponentially integrable), it is clear the inter-arrivals of τ^α are integrable.*

Going back to (3.14), we realize that $\sum_{k\in\mathbb{N}}(\Gamma^{*k})_{[0],[N]}(N)$ is a mass renewal function, explicitly:

$$\sum_{k\in\mathbb{N}} (\Gamma^{*k})_{[0],[N]}(N)\ =\ \mathbf{P}_{[0]}^\omega\left(N \in \tau^{[N]}\right). \qquad (3.16)$$

It is now clear that (3.14) generalizes (2.18).

Let us get to showing that $b = \mathrm{F}_\omega$: observe first that there exists $\mathrm{c} \in (0, 1)$ such that $\mathrm{c} \le \xi_\alpha/\xi_\beta \le 1/\mathrm{c}$ for every $\alpha, \beta \in \mathbb{S}$ and that of course $\mathbf{P}^{\omega}_{[0]}\left(N \in \tau^{[N]}\right) \leq 1$ for every N. So in order to establish the validity of (3.11) when $\lambda^{\omega}_{\star}(0) > 1$ it suffices to exhibit a lower bound on $\mathbf{P}^{\omega}_{[0]}\left(N \in \tau^{[N]}\right)$, even a very poor one. The only difficulty in establishing such a bound comes from the fact that the inter-arrival times are dependent: this problem is circumvented by considering for $\alpha \in \mathbb{S}$ the renewal $\tau^{\alpha} := \{\tau_j : J_j = \alpha\}$, see Remark 3.3. As we pointed out, $(\tau^{\alpha} - \tau^{\alpha}_1)/T$ is a positive recurrent renewal and this shows $c_0 := \inf_N \mathbf{P}^{\omega}_{[0]}\left(N \in \tau^{[N]}\right) > 0$, by the Renewal Theorem (Theorem A.3). We write explicitly the bounds that we have obtained

$$c_0 \mathrm{c} \leq \exp(-bN)Z^{\mathrm{c}}_{N,\omega} \leq 1/\mathrm{c}, \tag{3.17}$$

which is (a much stronger statement than) $b = \mathrm{F}_\omega$.

The case of $\lambda^{\omega}_{\star}(0) = 1$.

The steps of the previous case hold, if one sets $b = 0$, up to formula (3.16). Also the upper bound in (3.17) is still fine, although rough. The lower bound instead does not hold, since the inter-arrivals are not integrable. However there is no need of such a bound since we know *a priori* that $\mathrm{F}_\omega \geq 0$.

The case of $\lambda^{\omega}_{\star}(0) < 1$.

In this case we look directly at the matrix $\sum_n M_{\alpha,\beta}(n)$, which has Perron–Frobenius eigenvalue $\lambda^{\omega}_{\star}(0)$ and right eigenvector still denoted by ξ. On the other hand we define in this regime

$$\Gamma_{\alpha,\beta}(n) := M_{\alpha,\beta}(n)\frac{\xi_{\beta}}{\xi_{\alpha}}, \tag{3.18}$$

so that this time we have $\sum_{\beta,n} \Gamma_{\alpha,\beta}(n) < 1$, *i.e.* Γ is a sub-probability kernel, and formula (3.14) in this case is

$$Z^{\mathrm{c}}_{N,\omega} = \frac{\xi_{[0]}}{\xi_{[N]}} \sum_{k \in \mathbb{N}} (\Gamma^{k*})_{[0],[N]}(N). \tag{3.19}$$

Apart for the ratio of eigenvalues, the right-hand side is of course the mass renewal function of a terminating process, whose law is still denoted by $\mathbf{P}^{\omega}_{[0]}$. Some details on this process will be given in the next section. For the moment we are content with the fact that the right-hand side of (3.19) is bounded above by $\max_{\alpha,\beta} \xi_{\alpha}/\xi_{\beta}$, since this shows that $\mathrm{F}_\omega = 0$ (once again: we already know that $\mathrm{F}_\omega \geq 0$). $\qquad\square$

3.2 Sharp Estimates on the Partition Function

A byproduct of the proof of Theorem 3.2 is the formula

$$Z_{N,\omega}^{\mathrm{c}} = \exp(\mathrm{F}_\omega N) \frac{\xi_{[0]}}{\xi_{[N]}} \mathbf{P}_{[0]}^\omega \left(N \in \tau^{[N]} \right), \qquad (3.20)$$

that holds in all regimes. For the sake of conciseness we set

$$\delta := \lambda_\star^\omega(0). \qquad (3.21)$$

As we have already stressed, the analogy with (2.18) is evident and it is probably not surprising for the reader that from such a formula one can extract the sharp behavior of the partition function. It should however be noted that, while in the positive recurrent set-up ($\delta > 1$) the theory of Markov renewals is well developed, fewer results are available in the literature on the mass renewal function of null recurrent Markov renewals. Moreover the results, even only at the level of sharp asymptotic behavior of the partition function, are more involved. As we shall see, this complexity is not only of a technical nature, but it really reflects a substantially larger variety of phenomena that can be observed in weakly inhomogeneous models, with respect to homogeneous ones.

We have the following:

Theorem 3.4 *For every $\eta \in \mathbb{S}$ there exist positive constants $C_{\omega,\eta}^>$, $C_{\omega,\eta}^=$ and $C_{\omega,\eta}^<$ such that in the limit $N \to \infty$ with $[N] = \eta$ we have that*

(1) if $\delta > 1$ then $Z_{N,\omega}^{\mathrm{c}} \sim C_{\omega,\eta}^> \exp(\mathrm{F}_\omega N)$;
(2) if $\delta = 1$ then $Z_{N,\omega}^{\mathrm{c}} \sim C_{\omega,\eta}^= N^{-1/2}$;
(3) if $\delta < 1$ then $Z_{N,\omega}^{\mathrm{c}} \sim C_{\omega,\eta}^< N^{-3/2}$.

The quantities $C_{\omega,\eta}^>$, $C_{\omega,\eta}^=$ and $C_{\omega,\eta}^<$ have explicit expressions that are given below. Clearly this theorem generalizes Theorem 2.2 (of course in this cases we are considering only the recurrent case with $\alpha = 1/2$).

We will not give a full proof of Theorem 3.4, but we will stress on the ideas that lead to results. And case by case we will analyze the dependence of the constants on η since it has a direct impact on the behavior of the polymer trajectories.

Sharp estimates in the localized case ($\delta > 1$).

The generalization of Theorem A.3 to the Markov renewal setting is rather straightforward. Let us state this generalization in our context: recall that ξ is the right eigenvector of $A(b)$ with eigenvalue δ and call ζ the corresponding left eigenvector. We normalize ξ and ζ in such a way that $\sum_\alpha \zeta_\alpha \xi_\alpha = 1$, which leaves still a degree of freedom (which however does not affect the computations). In this way if we set $\nu_\alpha = \zeta_\alpha \xi_\alpha$ then ν is the invariant measure of the irreducible chain $\{J_k\}_k$. The transition matrix of this chain is of course $\sum_{n \in \mathbb{N}} \Gamma(n)$. Let us set

$$\mu := \sum_{\alpha,\beta \in S} \sum_{n \in \mathbb{N}} n \nu_\alpha \Gamma_{\alpha,\beta}(n) = \sum_{\alpha,\beta \in S} \sum_{n \in \mathbb{N}} n \exp(-F_\omega n) \zeta_\alpha M_{\alpha,\beta}(n) \xi_\beta \in (0,\infty).$$

$$(3.22)$$

Then

$$\lim_{\substack{n \to \infty \\ [n]=\beta}} \mathbf{P}_\alpha^\omega \left(n \in \tau^\beta \right) = T \frac{\nu_\beta}{\mu}, \qquad (3.23)$$

where τ^β has been defined in Rem. 3.3. From that remark and the classical Renewal Theorem one can build directly a proof of (3.23) (alternatively, see [Asmussen (2003), Theorem VII.4.3]).

From (3.23) and (3.20) we immediately obtain the statement in Theorem 3.4(1) with

$$C_{\omega,\eta}^> = C_{\omega,0,\eta}^> := \xi_0 \zeta_\eta \frac{T}{\mu}. \qquad (3.24)$$

Of course if we are considering a polymer chain between n and N instead of 0 and N the asymptotic behavior of $Z_{N-n,\theta^n\omega}^c$ changes in the obvious way.

Sharp estimates in the critical case ($\delta = 1$).

In this case there is no exponential growth and one is left with estimating the mass renewal function: the notation is the same as for $\delta > 1$, just keep in mind that now $b = 0$, so $A_{\alpha,\beta}^0(n) = M_{\alpha,\beta}(n)$, $\Gamma_{\alpha,\beta}(n) = M_{\alpha,\beta}(n)\xi_\beta/\xi_\alpha$ and $\sum_n \Gamma(n)$ is the Markov transition matrix of J, with invariant measure ν, $\nu_\alpha = \zeta_\alpha \xi_\alpha$. We will not give the details of the proof, that can be found in [Caravenna *et al.* (2005)], but we stress that the proof is a matter of dealing with a return distribution that is a random superposition of return laws with $\alpha = 1/2$ and trivial $L(\cdot)$, so the N dependence in Theorem 3.4(2) does not come as a surprise once we consider the corresponding result (2.15) for the homogeneous pinning model. The exact constant can be obtained by estimating the asymptotic behavior of the inter-arrival law $K_\alpha(\cdot)$ of τ^α, which is positive only on $T\mathbb{N}$ and on this set $K_\alpha(N) \sim N^{-3/2}(\zeta \cdot L\xi)/\nu_\alpha$ (of course $\cdot$ denotes the scalar product) with $L_{\alpha,\beta}$ the limit as $n \to \infty$, $[n] = \beta - \alpha$, of $n^{3/2}M_{\alpha,\beta}(n)$ which can be expressed explicitly:

$$L_{\alpha,\beta} = \frac{1}{2}c_K \exp(\omega_\beta)\left(1 + \mathbf{1}_{\mathrm{H}=0}\exp(\Sigma_{\alpha,\beta})\right). \tag{3.25}$$

Finally:

$$C^=_{\omega,\eta} = C^=_{\omega,0,\eta} := \frac{T^2\xi_0\zeta_\eta}{2\pi(\zeta \cdot L\xi)}. \tag{3.26}$$

Sharp estimates in the delocalized case ($\delta < 1$).

In this case Theorem 3.4(2) is the analog of (2.14): we have to estimate the mass renewal function of a transient process and it is not surprising that the decay of the transition probability is proportional to the decay of the mass renewal function, *cf.* Theorem A.4. Again we are left with establishing the precise asymptotic behavior and computing the constant and for this we refer to [Caravenna *et al.* (2005)]. We give here the value of the constant: let $B := \sum_n M(n)$ and recall that the Perron–Frobenius eigenvalue of B is δ. We have

$$C^<_{\omega,\eta} := \left((1 - B)^{-1}L(1 - B)^{-1}\right)_{0,\eta}, \tag{3.27}$$

and this formula can be interpreted by recognizing the contributions of a finite (but arbitrary) number of short jumps close to the left and right boundary of the system and a large jump covering almost all of $\{1, \ldots, N\}$.

3.3 The Limit Path Trajectories

We consider both infinite volume limits and scaling limits, but in both cases we just give the basic ideas of the proofs. We will then focus on the dependence of the results on the boundary conditions. In reality, the only cases in which there is a dependence of the limit path measure on the boundary conditions is when $\omega \in \mathcal{P}$:

$$\mathcal{P} := \{\omega : \delta < 1, \, \mathrm{H} = 0 \text{ and } \Sigma_{\alpha,\beta} \neq 0 \text{ for some } \alpha, \beta). \tag{3.28}$$

In other words, it is the case in which there is a non-trivial copolymer interaction, with centered charges, and the model is delocalized.

3.3.1 *The infinite volume limit*

We have the following result

Theorem 3.5 *With the exception of the case of $\omega \in \mathcal{P}$, both the polymer measure $\mathbf{P}^c_{N,\omega}$ and $\mathbf{P}^f_{N,\omega}$ converge as $N \to \infty$ to the same limit $\mathbf{P}_\omega$, distribution of an irreducible Markov chain on $\mathbb{Z}$. This Markov chain is:*

(1) positive recurrent if $\delta > 1$ (localized regime);
(2) null recurrent if $\delta = 1$ (critical regime);
(3) transient if $\delta < 1$ (delocalized regime).

If $\omega \in \mathcal{P}$ (and then $\delta < 1$) the sequence $\{\mathbf{P}^a_{N,\omega}\}_{N: [N]=\eta}$ converges to $\mathbf{P}^{a,\eta}_\omega$, law of an irreducible transient Markov chain on $\mathbb{Z}$.

As we shall see now, in all regimes $\mathbf{P}_\omega$ and $\mathbf{P}^{a,\eta}_\omega$ are explicitly constructed. The proof of this result follows from Theorem 3.4, we give only a partial proof, which however contains the main ideas.

Discussion and partial proof of Theorem 3.5. An important general fact is that we can view $\mathbf{P}^a_{N,\omega}$ in terms of

(1) the family τ of the return times to zero;
(2) the family of the excursions from zero to zero, that is $\{(S_{\tau_{k-1}+1}, \ldots, S_{\tau_k})\}_k$. In practice it is useful to consider separately the absolute value of the excursion, $e_k(n) = |S_{\tau_k+n}|$ for $n = 1, \ldots, \tau_k - \tau_{k-1}$, and s_k, which is (as usual) the sign of the excursion.

One sees directly that, conditionally on τ, the law of the absolute values $\{e_k\}_k$ of the excursions under $\mathbf{P}^a_{N,\omega}$ is the same as it would be under $\mathbf{P}$, in particular $\{e_k\}_k$ is a (conditionally) independent sequence. The signs of the excursions instead are affected, but in a way that is easily computed and conditional independence is preserved too: conditionally on τ the probability that $s_k = 1$ is equal to

$$1/\big(1 + \exp\big(-(\tau_k - \tau_{k-1})\mathrm{H} + \Sigma_{[\tau_{k-1}],[\tau_k]}\big)\big), \qquad (3.29)$$

if $k \leq \mathcal{N}_N(\tau)$. If $\tau_{k-1} < N$ but $\tau_k > N$ then one has to replace τ_k with N in (3.29). If instead $\tau_{k-1} \geq N$ the sign is just fair coin tossing (but we are not focusing on that side of the polymer chain).

It should then be clear that in studying the convergence of $\{\mathbf{P}^a_{N,\omega}\}_N$ it is sufficient to study the law of τ under $\mathbf{P}^a_{N,\omega}$. The rest of the process is then built as specified above by *attaching* excursions.

Since at the end, as we are going to see:

(1) the law of τ converges to the law of a Markov renewal;

(2) conditionally on τ the law of the excursions are independent and the law of the kth one is a function of τ_{k-1} and τ_k;

the limit process is what is usually referred to as a *Markov Additive Process* [Asmussen (2003)], that is a process that is the result of *adding contributions* whose law is decided by sampling an auxiliary Markov chain (in our case the chain J on $\mathbb{S}$).

Let us then consider the law of τ under $\mathbf{P}^a_{N,\omega}$: we write for $k \in \mathbb{N}$ and $n_0 := 0 < n_1 < \ldots < n_k \leq N$

$$\mathbf{P}^c_{N,\omega}\left(\tau_j = n_j \text{ for } j = 1, \ldots, k\right) =$$

$$\prod_{j=1}^{k} M_{[n_{j-1}],[n_j]}(n_j - n_{j-1}) \frac{Z^c_{N-n_k,\theta^{n_k}\omega}}{Z^c_{N,\omega}}. \quad (3.30)$$

Let us consider first the localized and the critical regimes ($\delta \geq 1$). By replacing $M_{\alpha,\beta}(n)$ with $\Gamma_{\alpha,\beta}(n)$ by using (3.12) and by applying Theorem 3.4 one gets

$$\lim_{N\to\infty} \mathbf{P}^c_{N,\omega}\left(\tau_j = n_j - n_{j-1} \text{ for } j = 1, \ldots, k\right) =$$

$$\frac{\xi_{[0]} C^q_{\omega,[n_k],[N]}}{\xi_{[n_k]} C^q_{\omega,[0],[N]}} \prod_{j=1}^{k} \Gamma_{[n_{j-1}],[n_j]}(n_j - n_{j-1}), \quad (3.31)$$

with q equal to $>$, if $\delta > 1$, or to $=$, if $\delta = 1$. By applying the explicit expressions (3.24) and (3.26) one sees that the pre-factor in the right-hand side of (3.31) is equal to 1. We have therefore shown that the limit random set of zeros is the Markov renewal with kernel $\Gamma(\cdot)$.

In the delocalized case, $\delta < 1$, the argument is absolutely analogous, and this time the limit Markov renewal is terminating, as clearly suggested by (3.18). However, the *constants* in general do not simplify and (3.18) gives the kernel of the limit Markov renewal if and only if $\omega \notin \mathcal{P}$. If $\omega \in \mathcal{P}$ the process depends on $[N]$. This issue is treated in detail [Caravenna *et al.* (2006b)] and (briefly) taken up again in Section 3.4 below.

We have chosen to consider only $\mathbf{P}^c_{N,\omega}$. The computations for $\mathbf{P}^f_{N,\omega}$ are absolutely analogous and the only technical point is to generalize Theorem 3.4. This is however straightforward (it has been done repeatedly in Chapter 2). □

3.3.2   The scaling limit

In this section we consider the law of the process $\{X_t^N\}_{t \in [0,1]}$ defined by $X_t^N := S_{Nt}/(\sigma N^{1/2})$ if $t \in \{0,1,\ldots,N\}/N$ and by linear interpolation if $t \in [0,1]\backslash(\{0,1,\ldots,N\}/N)$. Here $\sigma := \sqrt{\mathbf{E}[S_1^2]} = \sqrt{p}$. Let us call $\mathbf{Q}_{N,\omega}^a$ the law of $\{X_t^N\}_{t \in [0,1]}$. $\mathbf{Q}_{N,\omega}^a$ is a measure on $C^0([0,1];\mathbb{R})$ equipped with the uniform topology and we want to study the weak convergence of the sequence of measures $\{\mathbf{Q}_{N,\omega}^a\}_N$, that is the convergence in law of $\{X_\cdot^N\}_N$ (see Appendix A.1). For the sake of conciseness we restrict ourselves to the case of $\{\mathbf{Q}_{N,\omega}^c\}_N$ (full results can be found in [Caravenna *et al.* (2005)]).

We need some notations for Brownian processes on $[0,1]$ (these processes are treated in full detail in [Revuz and Yor (1999)]):

- $\mathbf{b} := \{\mathbf{b}_t\}_{t \in [0,1]}$ is the Brownian bridge;
- $\mathbf{e} := \{\mathbf{e}_t\}_{t \in [0,1]}$ is the Brownian bridge constrained to stay positive on $(0,1)$. This process is also called *normalized Brownian excursion* or Bessel Bridge of dimension 3 between 0 and 0. By $\mathbf{e}^{(p)} := \{\mathbf{e}_t^{(p)}\}_{t \in [0,1]}$ we mean $Y_p \mathbf{e}$, with Y_p a random variable taking values $+1$, with probability p, and -1 with probability $1 - p$ (Y_p and $\mathbf{e}$ are independent).
- $\mathbf{b}^{(p)} := \{\mathbf{b}_t^{(p)}\}_{t \in [0,1]}$ is the skew Brownian bridge of parameter $p \in (0,1)$, that is the Brownian bridge for which the sign of each excursion is chosen by independent (biased, if $p \neq 1/2$) coin tossing. In particular, $\mathbf{b}^{(1)} \sim |\mathbf{b}|$ and $\mathbf{b}^{(1/2)} \sim \mathbf{b}$.

We have the following:

Theorem 3.6   (Scaling limits).

(1) If $\delta > 1$ *(localized regime) then* $\{\mathbf{Q}_N^c\}_N$ *converges weakly to the measure concentrated on the constant function taking value zero.*

(2) If $\delta = 1$ *(critical regime) then* $\{\mathbf{Q}_N^c\}_N$ *converges weakly to the law of* $\mathbf{b}^{(p_\omega)}$, *with*

$$p_\omega := \frac{c_K \sum_\alpha \zeta_\alpha \sum_\beta \exp(\omega_\beta) \xi_\beta}{2\zeta \cdot L\xi}. \tag{3.32}$$

(3) If $\delta < 1$ *(delocalized regime) then, for every* $\eta \in \mathbb{S}$, $\{\mathbf{Q}_N^c\}_{N:\, [N]=\eta}$ *converges weakly to the law of* $\mathbf{e}^{(p_{\omega,\eta})}$, *with*

$$p_{\omega,\eta} := \frac{c_K \sum_{\alpha,\beta} (1-B)_{0,\alpha}^{-1} \exp(\omega_\beta)(1-B)_{\beta,\eta}^{-1}}{2\left((1-B)^{-1}L(1-B)^{-1}\right)_{0,\eta}}. \tag{3.33}$$

If $\omega \notin \mathcal{P}$ then $p_{\omega,\eta} = p_{\omega,\eta'}$ for every $\eta' \in \mathbb{S}$ and $\{\mathbf{Q}_N^c\}_N$ converges.

Let us briefly discuss the proof.

If $\delta > 1$ the proof follows easily as in the case of Theorem 2.5(1), applying of course the estimate in Theorem 3.4(1), and choosing $a = c$ or $a = f$ makes no difference. In particular one shows that (2.39) holds also in this weakly inhomogeneous context.

If $\delta < 1$ the argument of Theorem 2.5(2) applies: notice that it is just based on the fact that $N^{3/2} Z_{N,\omega}^c$ is bounded away from 0 and ∞ uniformly in N (in our case such an estimate is provided by Theorem 3.4(3), even if there is no need of sharp bounds so far). So for N large, with large probability the visits of the polymer to the interface are all very close to the boundary points, that is at a distance $o(N)$ (even $o(N^\varepsilon)$ for every $\varepsilon > 0$). So, except for a few short excursions close to the boundary, one simply observes a long random walk excursion. So there is no surprise that we observe only one Brownian excursion in the limit. However there is no reason for the sign of such an excursion to be deterministic: the distribution of the sign, that is the probability p_{ω_η} is computed directly, by using the sharp asymptotic estimate in Theorem 3.4(3).

If $\delta = 1$ we face first of all the problem of proving the convergence of the rescaled random set of zeros to $\{t : \mathsf{b}_t = 0\}$. This is of course analogous to Theorem 2.7(1). In this case however the successive returns to zero of the polymer are not close to a standard renewal, but to a Markov renewal. So one has first to overcome this difficulty and of course Theorem 3.4(2) is one of the ingredients. The rest of the argument exploits again the fact that the absolute value of the polymer path, conditionally on the returns, is just a collection of *random walk excursions*. Of course not all of these excursions can be approximated via Brownian excursions, since most of them are just of a finite length. However for any $\varepsilon > 0$ we can choose n such that the sum of the n longest excursions (we measure the length of an excursion in terms of the distance of the endpoints) is larger than $(1 - \varepsilon)N$ with probability larger than $1 - \varepsilon$ (this is quite clear if one thinks of the excursions of a Brownian bridge): with a different language, this argument has been explained also in Section 2.5. And these excursions can be approximated by Brownian excursions. Finally, with a lengthy computation one obtains the limit of the distribution of the signs of the excursions.

3.4   First Order Transitions, Non-Uniqueness and Phase Diagrams

The regime $\mathcal{P}$.

How can one understand the *anomalous* path behavior observed in $\mathcal{P}$? In order to give an answer to this question let us consider the particular case in which the energy is

$$\lambda \sum_{n=1}^{N} (\hat{\omega}_n + h) \operatorname{sign} \left((S_{n-1}, S_n)\right) - \widetilde{\lambda} \sum_{n=1}^{N} \delta_n, \qquad (3.34)$$

with $\lambda > 0$, $h \in \mathbb{R}$, $\hat{\omega}$ centered periodic and $\widetilde{\lambda} \in \mathbb{R}$. This is not the general case introduced in (3.1), since the pinning term is homogeneous, but, at a qualitative level, the phenomenon that we are considering is fully caught in this restricted set-up. The free energy is denoted by $f_\omega(\lambda, h, \widetilde{\lambda})$: this atypical notation for the free energy is due to the fact that this time $f_\omega(\lambda, h, \widetilde{\lambda}) \geq \lambda|h|$, as it is easily shown by considering only the trajectories spending all their time in the upper half plane, if $h > 0$, or in the lower half plane, if $h < 0$. In the case $h = 0$ we are in the classical set-up. Therefore in this case $\mathcal{L} = \{(\lambda, h, \widetilde{\lambda}) : f_\omega(\lambda, h, \widetilde{\lambda}) - \lambda|h| > 0\}$.

Let us give first the following result that is interesting in itself (we will not prove it here, see Section 3.5 for references):

Proposition 3.7   *For every $\lambda > 0$ we have $\mathrm{F}_\omega(\lambda, 0, 0) > 0$, therefore a copolymer with non-trivial centered periodic sequence of charges and zero external field is localized.*

Since $\mathrm{F}(\lambda, h, \cdot)$ is non-increasing, at $h = 0$ the polymer is therefore localized for every $\widetilde{\lambda} \leq 0$. However if $\widetilde{\lambda}$ is sufficiently large, that is when the depinning interaction is sufficiently strong, the polymer delocalizes and therefore (by monotonicity) there exists $\widetilde{\lambda}_c(\lambda, h)$, with $\widetilde{\lambda}_c(\lambda, 0) > 0$, such that $f(\lambda, h, \widetilde{\lambda}) = 0$ for $\widetilde{\lambda} \geq \widetilde{\lambda}_c(\lambda, h)$. This is seen by observing first that, by the monotonicity properties of $h \mapsto f(\lambda, h, \widetilde{\lambda}) - \lambda|h|$ (monotonic non-increasing for $h \geq 0$ and non-decreasing for $h \leq 0$), it is sufficient to show that $f(\lambda, 0, \widetilde{\lambda}) = 0$ for $\widetilde{\lambda}$ sufficiently large. However the maximal energetic contribution coming from an excursion from the $\hat{\omega}$ charges is bounded, that is $q := \sup_{n,k} |\hat{\omega}(n, n+k)| < \infty$ because it is the maximum over a finite number of possibilities, and $f(\lambda, 0, \widetilde{\lambda}) = 0$ if $\widetilde{\lambda} \geq q$ and the claim is proven.

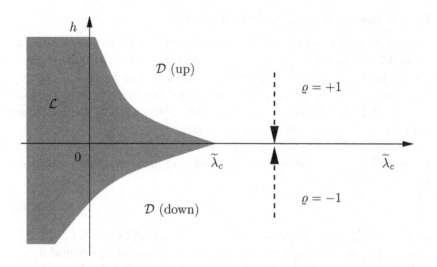

Fig. 3.1 The (qualitative) phase diagram of a periodic copolymer with pinning strength $-\lambda$, that is the model with energy given in (3.34). The system with parameters chosen above the half-line $(\widetilde{\lambda}_c, \infty)$ has $\varrho := \lim_{N \to \infty} (1/N) \sum_{n=1}^{N} \mathbf{P}^a_{\omega, N}(s_n = 1) = +1$, while it has $\varrho = -1$ below. Co-existence of the delocalized up and down phases is observed for $h = 0$ and $\widetilde{\lambda} > \widetilde{\lambda}_c$.

Let us choose now $\widetilde{\lambda} \geq \widetilde{\lambda}_c(\lambda, 0)$. The system is delocalized, with *a priori* no evident preference for the upper or the lower half plane. But if we switch on a positive external field, that is $h > 0$, the system is delocalized above: one way to see this is to take the derivative in h of the free energy that in this regime is just $\lambda|h|$. And of course it is delocalized below if h is negative. In particular for the choice of λ and $\widetilde{\lambda}$ that we have made $f_\omega(\lambda, \cdot, \widetilde{\lambda})$ is not differentiable in 0, in fact the derivative jumps from $-\lambda$ to $+\lambda$. It comes as no surprise then that for $h = 0$ there is a phenomenon of phase coexistence: the limit $h \searrow 0$ of the delocalized measure obtained for $h > 0$ and the one obtained for $h \nearrow 0$ are clearly different (one is delocalized above and the other is delocalized below the interface). It is therefore no longer surprising that taking limits at $h = 0$ along different subsequences one can end up on limit points that are different linear combinations of the two measures we have obtained above. This is indeed what happens [Caravenna *et al.* (2006b)], but showing it is rather delicate and lengthy and formulating the complete result would involve notions from the general theory of Gibbs measures [Georgii (1988)].

Periodic pinning versus homogeneous pinning.

We include in this section also another result that is in the spirit of Proposition 3.7. Consider the general model (3.1), but set $\hat{\omega}_n = 0$ for every n, that is there is no copolymer interaction (we denote by F_ω its free energy). The charges ω are instead periodic but not necessarily centered. The result we are going to state says that the periodic pinning model *localizes more* than the corresponding homogeneous model.

Proposition 3.8   *For every non-trivial, i.e. non-constant, periodic sequence ω we have*

$$F_\omega > F(h), \tag{3.35}$$

with $h := \sum_{n=1}^{T(\omega)} \omega_n / T(\omega)$, if h is in the closure of $\{h : F(h) > 0\}$, i.e. (3.35) holds for values of h corresponding to the localized and critical regime of the homogeneous model, and therefore (by continuity) also for some h in the delocalized regime.

Proof. Let us redefine ω_n as $\omega_n + h$, so that ω is centered, and let us consider the model in which ω is replaced by $t\omega$, $t \in \mathbb{R}$. Going back to Theorem 3.2 and to formula (3.10) we are interested in computing the Perron–Frobenius eigenvalue of the matrix with positive entries

$$B_{\alpha,\beta}(t,h;b) := \sum_n \exp\left(t\omega_\beta + h\right) \exp(-bn) K(n) \mathbf{1}_{n \in \beta - \alpha}, \tag{3.36}$$

where $b \geq 0$. Below we often omit the dependence on b. The map $t \mapsto \lambda_\star(B_{\alpha,\beta}(t,h))$ is log-convex (in particular convex) and analytic (see Section A.8). Now we observe that $B_{\alpha,\beta}(0,h) = \exp(h) W_{\beta-\alpha}$, with $W_\eta := \sum_{n \in \eta} \exp(-bn) K(n)$. The vector $u := (1,\ldots,1)$ is both right and left eigenvector of the matrix $W := \{W_{\beta-\alpha}\}_{\alpha,\beta}$ with eigenvalue, in fact the Perron–Frobenius eigenvalue, equal to $\sum_n \exp(-bn) K(n)$. From this we have (see Appendix A.8)

$$\frac{d}{dt}\lambda_\star(B_{\alpha,\beta}(t,h))\Big|_{t=0} = u \cdot \left(\frac{d}{dt}B(t,h)\Big|_{t=0}\right) u = \sum_{\alpha,\beta} e^h W_{\beta-\alpha}\omega_\beta = 0, \tag{3.37}$$

where in the last step we have used $\sum_\beta \omega_\beta = 0$. Since the function $\lambda_\star(B_{\alpha,\beta}(\cdot,h))$ is convex, by (3.37) it has a minimum at the origin. But $\lambda_\star(B_{\alpha,\beta}(\cdot,h))$ is also analytic and therefore it is equal to the minimum in an interval only if it is constant on all its domain of definition, that is $\mathbb{R}$.

But $\lim_{t\to\infty}\sum_{\alpha,\beta}B_{\alpha,\beta}(t,h)=\infty$ which implies $\lim_{t\to\infty}\lambda_*(B(t,h))=\infty$ and therefore $\lambda_*(B(t,h))>\lambda_*(B(0,h))$ if $t\neq 0$. Now choose $b\geq 0$ such that $\lambda_*(B(0,h;b))=1$, that is $b=\mathrm{F}(h)$. Therefore for any $t\neq 0$ we have $\lambda_*(B(t,h;b))>1$, in particular for $t=1$. This means that $\lambda_*(B(1,h;\widetilde{b}))=1$ implies $\widetilde{b}>\mathrm{F}(h)$ and $\widetilde{b}$ is precisely F_ω. $\qquad\square$

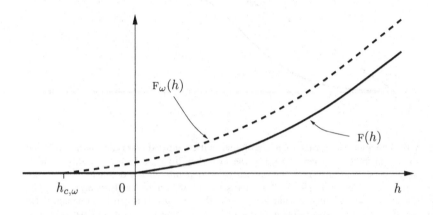

Fig. 3.2   The free energy of the periodic pinning model compared to the one of the associated homogeneous model.

The periodic copolymer.

Consider the model (3.34) with $\widetilde{\lambda}=0$. This is the copolymer model *without* adsorption term. However, as we already pointed out for example in Remark 1.11, we are fully entitled to talk about absence of adsorption term if the interface is really neutral and this has to do with the definition of $\mathrm{sign}((S_{n-1},S_n))$ when $(S_{n-1},S_n)=0$. But before talking about neutrality, let us point out that by soft, mainly convexity, arguments one can show that there exists a critical line, that is in fact a continuous function $h_c:[0,\infty)\to[0,\infty)$, separating $\mathcal{L}$ and $\mathcal{D}$, see Figure 3.3. We will not give these arguments here since they are given in detail in Chapter 6, see in particular Theorem 6.1, for the disordered case, but we signal that what *a priori* is not obvious is that $h_c(\lambda)>0$ for every $\lambda>0$, which means in particular that if $h=0$ the copolymer is localized no matter how weak the interaction is. Note that, since the free energy is non-decreasing in λ (this follows immediately from the convexity in λ and the fact that the free energy is non-negative), it is sufficient to study the case of λ small to infer localization for arbitrary $\lambda > 0$.

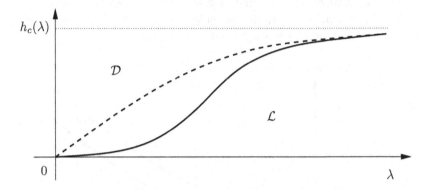

Fig. 3.3   The phase diagram of a copolymer with centered periodic charges. Some features are absolutely general, notably the fact that $\lim_{\lambda \to \infty} h_c(\lambda) > 0$ exists and it is at most equal to $-\min_n \hat{\omega}_n$: the precise value depends on the definition of the sign assigned to a monomer lying on the interface, but it can be easily computed. The behavior at the origin is instead harder to estimate, but also this can be done explicitly and in *full* absence of adsorption $h_c(\lambda) \sim c_\omega \lambda^3$ for λ small, c_ω is a positive constant, while in presence of adsorption (dashed line) the slope at the origin is positive.

And it is precisely in the small λ behavior of the free energy and of the critical curve that the issue of the neutrality of the interface comes up. Intuitively, if only the value of the charges and their location with respect to the interface, above or below, matter, in a symmetric situation $(h = 0)$ an excursion of a fixed portion of the polymer above the interface should yield an interaction energy equal to minus the interaction energy for an excursion below. This leads to an energy contribution equal to $\log \cosh(\lambda \hat{\omega}(n_1, n_2])$ for every excursion $\{n_1, \ldots, n_2\}$, that is for every $n_1 < n_2$. For small λ this leads to an energy $O(\lambda^2)$ per excursion and a pinning model based on random walk, see *e.g.* Section 1.2.1, yields a free energy $O(\lambda^4)$. Since the asymmetry enters like λh, one guesses that $h_c(\lambda)$ should behave like λ^3 for small λ. This is indeed correct, and it happens for example when one chooses $\text{sign}((S_{n-1}, S_n))$ by flipping a fair coin when $(S_{n-1}, S_n) = (0, 0)$. The choice that we have made in this section, that is -1, exits from the notion of neutrality given above and in fact for excursions of length one the energetic contribution is proportional to λ and therefore much larger. This corresponds to a pinning effect that suffices to lead to a positive slope for $h_c(\cdot)$ at the origin.

3.5 Bibliographic Complements

Complements to Section 3.1

The formula for the free energy based on the matrix encoding procedure has been derived in [Bolthausen and Giacomin (2005)] by Large Deviations arguments: the periodic Hamiltonian can be written as an empirical functional of the return times to zero of the random walk and the free energy can be then expressed in terms of a variational problem that can be solved. The renewal theory approach, developed in [Caravenna *et al.* (2005)] and in [Caravenna *et al.* (2006b)], that we present here is instead more elementary, but less intuitive and it has been found from the structure of the solution itself. It should be stressed however that the renewal theory approach goes naturally well beyond Large Deviations.

A lot can be found in the literature on periodic models, see *e.g.* [Galluccio and Graber (1996)], [Garel *et al.* (2000)], [Naidedov and S. Nechaev (2001)], [Nechaev and Zhang (1995)], [Rechnitzer and Janse van Rensburg (2004)], [Sinai and Spohn (1996)], [Sommer and Daoud (1995)], [Sommer and Daoud (1996)] and [Whittington (1998)], and there are at least two reasons for this:

(1) In theoretical physics periodic models find motivation as caricatures of disordered models: they capture some of the inhomogeneity of disordered models, but they still retain the *solvable* character of homogeneous models.
(2) Periodic models, notably the so called *block copolymers* are of central importance in application, for example in material science and in biology. We omit the extensive bibliography since it is too far from the mathematical aspects on which we focus.

It should be however noticed that for much of the literature the focus is on the case $T = 2$, which in fact can be reduced to a homogeneous case by one *decimation step*: in mathematical terms this just corresponds to looking at the marginal on (say) even sites. In principle this approach can be taken also with period T, but in practice it turns out to be an extremely cumbersome procedure as T becomes large and the arising model, which for $T = 2$ is precisely a pinning model (see Chapter 2), for larger T is more complex (and in fact, as we have seen, the phenomenology is richer). The works of Sommer and Daoud mentioned above deal with the general T case, but only from a non rigorous viewpoint (obtaining the λ^3-phenomenology explained in Figure 3.3, see [Bolthausen and Giacomin (2005)] for rigorous results). In [Rechnitzer and Janse van Rensburg (2004)] some very special sequences on periodic charges, with $T > 2$, are treated.

Complements to Section 3.2

The results of this section are taken from [Caravenna *et al.* (2005)].

Complements to Section 3.3

The results of this section are taken from [Caravenna *et al.* (2005)] (scaling limits) and from [Caravenna *et al.* (2006b)] (infinite volume limits).

Complements to Section 3.4

In [Bolthausen and Giacomin (2005)] one can find the small λ analysis of the phase diagram, leading to $h_c(\lambda) \sim c\lambda^3$, with explicit $c > 0$. The analysis is restricted to copolymers based on simple random walk, but it extends to the cases treated here if one takes into account the remarks on neutrality of the interface (note for example that the interface for the copolymer model in [Garel *et al.* (2000)] is not neutral and the positive slope phenomenon is found). This result implies Proposition 3.7, however that argument has to be adapted to our more general set-up: a general (and different) proof can be found [Caravenna *et al.* (2005)]. More details on the first order transition issue for copolymers with depinning can be found in [Caravenna *et al.* (2006b)].

Chapter 4

The Free Energy of Disordered Polymer Chains

4.1 Preliminary Observations

In this chapter we work exclusively in the general return time set-up, *cf.* Section 1.8 and its notations, in particular the measures we consider have Boltzmann factors containing the energy (1.75). Recall that we are assuming that the functions $\varphi_{(i)}(\cdot)$, $i = 1, 2, 3$, in (1.76) and in (1.77) have sub-linear growth and they are continuous. Moreover $\varphi_{(2)}(0) = 0$ and (1.78) holds. We observe that

$$\mathcal{H}_\omega^{\text{exc}}(i, k) = \mathcal{H}_{\theta^i \omega}^{\text{exc}}(0, k - i), \tag{4.1}$$

and the same property holds for $\mathcal{H}_\omega^{\text{i-exc}}(i, k)$. One moreover verifies directly that there exists $C > 0$ such that the following list of properties and bounds and on $\mathcal{H}_{N,\omega}(\tau)$, $\mathcal{H}_\omega^{\text{exc}}(i, k)$ and $\mathcal{H}_\omega^{\text{i-exc}}(i, k)$ hold.

(1) *Sub-linear growth of the energy.* For $s = \text{exc}$ and $s = \text{i-exc}$ we have

$$\left| \mathcal{H}_\omega^s(i, k) \right| \leq C \sum_{n=i+1}^{k} \left(|\omega_n| + |\hat{\omega}_n| \right), \tag{4.2}$$

which imply the same bound on $\sup_\tau |\mathcal{H}_{N,\omega}(\tau)|$, with $i = 0$ and $k = N$. Note that these are very rough bounds. Since $\varphi_{(2)}(\cdot)$ is bounded below we have also

$$\mathcal{H}_\omega^{\text{exc}}(i, k) \geq -C(1 + |\omega_k|). \tag{4.3}$$

(2) *Additive property under cutting at a renewal point.* If $M \leq N$ and $M \in \tau$ then

$$\mathcal{H}_{N,\omega}(\tau) = \mathcal{H}_{M,\omega}(\tau) + \mathcal{H}_{N-M,\theta^M \omega}(\theta^M \tau). \tag{4.4}$$

$\mathcal{H}_{N-M,\theta^M\omega}(\theta^M\tau)$ will also be denoted $\mathcal{H}_{M,N,\omega}(\tau)$.

(3) *Approximate sub-additive property under cutting in between inter-arrivals.* For $k > i+1$ and $m \in \{i+1,\ldots,k-1\}$

$$\mathcal{H}_\omega^{\text{exc}}(i,k) \le C(1+|\omega_m|) + \mathcal{H}_\omega^{\text{exc}}(i,m) + \mathcal{H}_\omega^{\text{exc}}(m,k). \qquad (4.5)$$

For this inequality we have used explicitly (1.78) and ω_m enters because the extra contact brings along extra energy.

(4) *Completing the incomplete excursion.* Since an incomplete excursion just lacks a contact, we have

$$\mathcal{H}_\omega^{\text{i-exc}}(m,k) = \mathcal{H}_\omega^{\text{exc}}(m,k) - \varphi_{(1)}(\omega_k), \qquad (4.6)$$

and $|\varphi_{(1)}(\omega_k)| \le C(1+|\omega_k|)$.

4.2   Existence of the Free Energy and Self-Averaging

This section is devoted to the proof of the *existence of the quenched free energy.* The free energy of a disordered system is *a priori* a random variable, however we will show that it is a degenerate random variable, *i.e.* a constant: this phenomenon is called *self-averaging* property of the free energy.

Theorem 4.1   *Let us assume that ω is IID (in the sense of Definition 1.16) and ω_1 and $\hat{\omega} \in L^1$. The limit of the sequence $\{N^{-1}\log Z_{N,\omega}^a\}_N$ exists $\mathbb{P}(\,d\omega)$-a.s. and in L^1, both for $a = \text{c}$ and $a = \text{f}$. Moreover there exists a real number F such that*

$$\text{F} = \lim_{N\to\infty} \frac{1}{N}\log Z_{N,\omega}^{\text{c}} = \lim_{N\to\infty} \frac{1}{N}\log Z_{N,\omega}^{\text{f}}, \qquad \mathbb{P}(\,d\omega)\text{-a.s..} \qquad (4.7)$$

The first step in the proof is controlling the so-called *quenched averaged free energy*: by the bound on $\mathcal{H}_{N,\omega}(\tau)$ in point (1) in Section 4.1, and by the hypothesis on ω, $\log Z_{N,\omega}^{\text{c}} \in L^1$. As a matter of fact the same bound implies directly

$$\sup_N \frac{1}{N}\mathbb{E}\log Z_{N,\omega}^{\text{c}} < \infty. \qquad (4.8)$$

So that the sequence with which we wish to define, in the limit, the quenched averaged free energy is bounded above. By using the same estimates we see also that the sequence is bounded below, but this is implicit in the following elementary and important observation:

Proposition 4.2 *The sequence $\{\mathbb{E}\log Z^{\text{c}}_{N,\omega}\}_N$ is super-additive.*

Proof. Set $a_N := \mathbb{E}\log Z^{\text{c}}_{N,\omega}$. Since $Z^{\text{c}}_{0,\omega} = 1$, we may assume that $N \geq 2$ and it is sufficient to show that $a_N \geq a_M + a_{N-M}$ for every $M \in \{1,\ldots,N-1\}$. Since

$$\mathbb{E}\left[\exp\left(\mathcal{H}_{0,M,\omega}(\tau)\right)\mathbf{1}_{M\in\tau}\exp\left(\mathcal{H}_{M,N,\omega}(\tau)\right)\mathbf{1}_{N\in\tau}\right] = Z^{\text{c}}_{0,M,\omega}Z^{\text{c}}_{M,N,\omega}, \quad (4.9)$$

and since the left-hand side is smaller (recall (4.4)) than $Z^{\text{c}}_{0,N,\omega}$ we have

$$\log Z^{\text{c}}_{0,N,\omega} \geq \log Z^{\text{c}}_{0,M,\omega} + \log Z^{\text{c}}_{M,N,\omega}. \quad (4.10)$$

By taking the $\mathbb{P}$-expectation, keeping in mind that $\log Z^{\text{c}}_{M,N,\omega}$ and $\log Z^{\text{c}}_{0,N-M,\omega}$ coincide in law, we obtain the statement. $\qquad\square$

Combining Proposition 4.2 and Proposition A.12 the existence of the limit of the sequence $\{\mathbb{E}\log Z^{\text{c}}_{N,\omega}\}_N$ is established and we set

$$\mathrm{F} := \lim_{N\to\infty}\frac{1}{N}\mathbb{E}\log Z^{\text{c}}_{N,\omega}\left(= \sup_N \frac{1}{N}\mathbb{E}\log Z^{\text{c}}_{N,\omega}\right). \quad (4.11)$$

Note that the use of F is consistent with the statement of Theorem 4.1. By (4.8), F is finite.

Lemma 4.3 *Under the hypotheses of Theorem 4.1*

$$\liminf_{N\to\infty}\frac{1}{N}\log Z^{\text{c}}_{N,\omega} \geq \mathrm{F}, \qquad \mathbb{P}(\,\mathrm{d}\omega) - a.s.. \quad (4.12)$$

Proof. This is just a consequence of (4.10) and of the Strong Law of Large Numbers (Theorem A.1). Fix $M \in \mathbb{N}$ and write $N = jM + K$, $j \in \mathbb{N}\cup\{0\}$ and $K \in \{0,\ldots,M-1\}$. By iterated application of (4.10) we have

$$\frac{1}{N}\log Z^{\text{c}}_{N,\omega} \geq \frac{M}{N}\sum_{i=0}^{j-1}\frac{1}{M}\log Z^{\text{c}}_{M,\theta^{iM}\omega} + \frac{1}{N}\log Z^{\text{c}}_{K,\theta^{jM}\omega}. \quad (4.13)$$

Since M is fixed, K/N vanishes in the limit $N \to \infty$. This implies of course that $M/N \to 1/j$, but also that the last term in the right-hand side of (4.13) is $\mathbb{P}(\,\mathrm{d}\omega)$-a.s. negligible. This is simply a consequence of the fact that $X(\omega) := \max_{K=0,\ldots,M-1}\left|\log Z^{\text{c}}_{K,\omega}\right|$ is a random variable in L^1 and that the sequence $\left\{ X\left(\theta^{jM}\omega\right)\right\}_j$ is IID, so that $X\left(\theta^{jM}\omega\right)/j$ vanishes a.s. as $j \to \infty$, see (A.1).

We are therefore left with the first term in the right-hand side of (4.13): also $\left\{ (1/M) \log Z^{\mathrm{c}}_{M,\theta^{iM}\omega}\right\}_i$ is a sequence of IID integrable random variables. By the Strong Law of Large Numbers, from (4.13) we conclude that a.s.

$$\liminf_{N\to\infty} \frac{1}{N} \log Z^{\mathrm{c}}_{N,\omega} \geq \frac{1}{M} \mathbb{E} \log Z^{\mathrm{c}}_{M,\omega}. \tag{4.14}$$

Since M is arbitrary, we can take the supremum (or the limit) in M and the proof is complete. $\square$

We now give a last lemma, that complements Lemma 4.3.

Lemma 4.4   *Under the hypothesis of Theorem 4.1*

$$\limsup_{N\to\infty} \frac{1}{N} \log Z^{\mathrm{c}}_{N,\omega} \leq \mathrm{F}, \qquad \mathbb{P}(\,\mathrm{d}\omega) - a.s.. \tag{4.15}$$

Proof. The crucial point is obtaining a result analogous to (4.10): we claim that there exists a constant $c > 0$ such that for every $M \in \{1,\dots,N-1\}$ and every ω

$$\log Z^{\mathrm{c}}_{0,N,\omega} \leq \log Z^{\mathrm{c}}_{0,M,\omega} + \log Z^{\mathrm{c}}_{M,N,\omega} + \log\left(c(M\wedge(N-M))^c\right) + C\left|\omega_M\right|. \tag{4.16}$$

(C is the constant introduced in Section 4.1.) By writing once again N as $jM + K$ and by applying j times (4.16) one obtains

$$\frac{1}{N} \log Z^{\mathrm{c}}_{N,\omega} \leq \frac{M}{N} \sum_{i=0}^{j-1} \frac{1}{M} \log Z^{\mathrm{c}}_{M,\theta^{iM}\omega} + \frac{1}{N} \log Z^{\mathrm{c}}_{K,\theta^{jM}\omega}$$

$$+ \frac{j}{N} \log\left(cM^c\right) + C\frac{j}{N}\frac{1}{j}\sum_{i=1}^{j}\left|\omega_{iM}\right|. \tag{4.17}$$

By the Strong Law of Large Numbers we obtain that a.s.

$$\limsup_{N\to\infty} \frac{1}{N} \log Z^{\mathrm{c}}_{N,\omega} \leq \frac{1}{M}\mathbb{E}\log Z^{\mathrm{c}}_{M,\omega} + \frac{1}{M}\log\left(cM^c\right) + C\frac{1}{M}\mathbb{E}\left|\omega_M\right|, \tag{4.18}$$

and the result follows by taking $M \to \infty$.

We are therefore left with giving a proof of (4.16). For this purpose we use the decomposition

$$Z_{0,N,\omega}^c = Z_{0,M,\omega}^c Z_{M,N,\omega}^c + \sum_{\substack{j,k:\,0\le j<M \\ M<k\le N}} Z_{0,N,\omega}^c(E_{j,k}), \qquad (4.19)$$

with $E_{j,k} := \{\tau : \tau_{\mathcal{N}_M(\tau)} = j,\ \tau_{\mathcal{N}_M(\tau)+1} = k\}$, that is $E_{j,k}$ is the event that the closest point in τ to the left of M is j and to the right is k. We write

$$Z_{0,N,\omega}^c(E_{j,k}) = Z_{0,j,\omega}^c K(k-j)\exp\left(\mathcal{H}_\omega^{\mathrm{exc}}(j,k)\right) Z_{k,N,\omega}^c. \qquad (4.20)$$

We now observe that by (4.5)

$$\exp\left(\mathcal{H}_\omega^{\mathrm{exc}}(j,k)\right) \le \exp(C(1+|\omega_M|))\exp\left(\mathcal{H}_\omega^{\mathrm{exc}}(j,M)\right)\exp\left(\mathcal{H}_\omega^{\mathrm{exc}}(M,k)\right), \qquad (4.21)$$

and by Definition 1.4 for every $\alpha_+ > \alpha$ there exists a positive constant c_1 such that

$$\frac{K(k-j)}{K(k-M)K(M-j)} \le c_1\left((k-M)\wedge(M-j)\right)^{1+\alpha_+} \\ \le c_1\left((N-M)\wedge M\right)^{1+\alpha_+}, \qquad (4.22)$$

for all admissible j and k. By using (4.21) and (4.22) in (4.20) we obtain that for some $c_2 > 0$

$$\sum_{\substack{j,k:\,0\le j<M \\ M<k\le N}} Z_{0,N,\omega}^c(E_{j,k}) \le c_2\exp\left(C|\omega_M|\right) N^{1+\alpha_+} Z_{0,M,\omega}^c Z_{M,N,\omega}^c. \qquad (4.23)$$

Inserting this estimate into (4.19) completes the proof. □

Proof of Theorem 4.1. Lemma 4.3 and Lemma 4.4 already guarantee the existence of the a.s. limit of the sequence $\{N^{-1}\log Z_{N,\omega}^c\}_N$ and that this limit takes the value F, defined in (4.11). By (4.3) and by restricting the **P** integral to the event $\{\tau_1 = N\}$ we obtain

$$\frac{1}{N}\log Z_{N,\omega}^c \ge \frac{1}{N}\log K(N) - \frac{C}{N}(1+|\omega_N|) := \xi_N, \qquad (4.24)$$

and $\xi_N \overset{N\to\infty}{\to} 0$ both almost surely and in L^1 (we have applied (A.1)). Therefore the random sequence $\{N^{-1}\log Z_{N,\omega}^c - \xi_N\}_N$ of non-negative random variables converges (a.s.) to F, as well as the deterministic sequence $\{N^{-1}\mathbb{E}\log Z_{N,\omega}^c - \mathbb{E}\xi_N\}_N$. By Schéffé's Lemma ([Williams (1991), p. 55]) the random sequence converges also in L^1, so that $\lim_{N \to \infty} N^{-1} \log Z^c_{N,\omega} =$ F also in the L^1 sense.

We are left with extending these convergence results to the free case. The result is a direct consequence of the bounds

$$Z^c_{N,\omega} \leq Z^f_{N,\omega} \leq C_1 N \exp(C|\omega_N|) Z^c_{N,\omega}, \tag{4.25}$$

where C_1 is a suitable constant.

The proof of (4.25) follows from the formula

$$Z^f_{N,\omega} = Z^c_{N,\omega} + \sum_{n=1}^{N} Z^c_{N-n,\omega} \exp\left(\mathcal{H}^{\text{i-exc}}_\omega(N-n, N)\right) \sum_{j>n} K(j). \tag{4.26}$$

In particular the lower bound is immediate. For the upper bound we observe that completing an incomplete excursion leads to a change in the energy that is due only to the last contact, *cf.* (4.6). Moreover the entropic factor changes, but since $\sum_{j>n} K(j) \overset{n \to \infty}{\sim} L(n)/(\alpha n^\alpha)$, there exists a constant $c > 0$ such that $\sum_{j>n} K(j) \leq cnK(n) \leq cNK(n)$. Overall:

$$\sum_{n=1}^{N} Z^c_{N-n,\omega} \exp\left(\mathcal{H}^{\text{i-exc}}_\omega(N-n, N)\right) \sum_{j>n} K(j)$$

$$\leq cN \exp\left(C(1 + |\omega_N|)\right) \sum_{n=1}^{N} Z^c_{N-n,\omega} \exp\left(\mathcal{H}^{\text{exc}}_\omega(N-n, N)\right) K(n)$$

$$= cN \exp\left(C(1 + |\omega_N|)\right) Z^c_{N,\omega}, \tag{4.27}$$

and from this and (4.26), the proof of (4.25) follows.   Theorem 4.1
                                                                  $\square$

4.3   Periodic versus Disordered Charge Sequences

Given the (random or deterministic) charges ω, we call $[\omega]_T$ the family of charges of period $T \in \mathbb{N}$ such that $([\omega]_T)_n = \omega_n$ and $([\hat{\omega}]_T)_n = \hat{\omega}_n$ for $n = 1, \ldots, T$.

The arguments of the previous section yield a straightforward proof of the convergence of $F_{[\omega]_T}$ toward F, when ω is a typical realization of the random charges ω.

Theorem 4.5   *Under the same assumptions as in Theorem 4.1, we have* *that*

$$\lim_{T \to \infty} F_{[\omega]_T} = F, \qquad (4.28)$$

$\mathbb{P}(d\omega)$–*a.s. and in the* $L^1(\Omega, \mathcal{A}, \mathbb{P})$ *sense.*

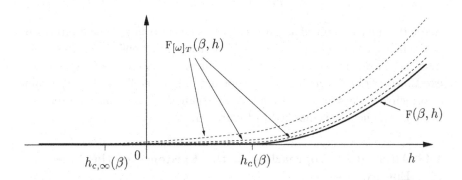

Fig. 4.1 A (possible) plot of the free energy for a disordered pinning model (thick line) and for three approximating periodic models, at β fixed. Even if $\lim_{T \to \infty} F_{[\omega]_T}(\beta, h) = F(\beta, h)$, what could happen is for example that $\lim_{T \to \infty} h_{c,[\omega]_T}(\beta) = h_{c,\infty}(\beta) < h_c(\beta)$, as it is shown in the figure.

Proof. We use the *chopping* procedure used in proving Lemma 4.3 and Lemma 4.4. In fact a direct byproduct of the proofs of these two lemmas is that for every n

$$\frac{1}{n} \sum_{j=0}^{n-1} X_j^-(\omega) \le \frac{1}{N} \log Z_{N,\omega}^c \le \frac{1}{n} \sum_{j=0}^{n-1} X_j^+(\omega), \qquad (4.29)$$

where $N = nT$ where $\left\{ X_j^{\pm} \right\}_j$ are two sequences of functions of ω. More precisely $X_j^-(\omega) := T^{-1} \log Z_{T, \theta^{jT}\omega}^c$ and $X_j^+(\omega) := X_j^-(\omega) + T^{-1} c \left(\log(T) + |\omega_{jT}| \right)$ for some $c > 0$ that does not depend on the choice of ω.

Replace ω with $[\omega]_T$ in (4.29) and $X_j^{\pm}(\cdot) = X_0^{\pm}(\cdot)$ for every j. By taking the limit $N \to \infty$ we obtain for every ω

$$0 \le F_{[\omega]_T} - \frac{1}{T} \log Z_{T,\omega}^c \le \frac{c}{T} \left(\log(T) + |\omega_T| \right). \qquad (4.30)$$

The rightmost term in this formula vanishes $\mathbb{P}(d\omega)$–a.s. and in L^1 as T tends to infinity. Therefore, by Theorem 4.1, the proof is complete. □

With self-explaining notation, Theorem 4.5 may suggest that $\{\mathcal{L}_{[\omega]_T}\}_T$ (and therefore $\{\mathcal{D}_{[\omega]_T}\}_T$) converges in a suitable sense to $\mathcal{L}$, respectively to $\mathcal{D}$. However, if $\underline{v} \in \mathcal{L}$, then $\mathrm{F}(\underline{v}) > 0$ and, by Theorem 4.5, $\mathrm{F}_{[\omega]_T}(\underline{v}) \geq \mathrm{F}(\underline{v})/2 > 0$ for T larger than a certain $n(\omega)$, $n(\omega) < \infty$ a.s.. This means that a.s.

$$\liminf_{T \to \infty} \mathcal{L}_{[\omega]_T} \supset \mathcal{L}, \tag{4.31}$$

where the inferior limit of a sequence of sets has of course to be interpreted as the inferior limit of the sequence of indicator functions. One cannot extract more from Theorem 4.5 and the problem does not come from possible *irregularities* of $\partial \mathcal{L}$, but it is more substantial and the limit of $\mathcal{L}_{[\omega]_T}$ could in principle be a substantially larger set than $\mathcal{L}$. What can go wrong in this convergence is exemplified in Figure 4.1.

4.4　Alternative Approaches to the Existence of the Free Energy

What we have presented in Section 4.2 is a proof of the existence of the free energy *by bare hands*, based on super-additivity of numerical sequences and on the Law of Large Numbers. It is a useful exercise in manipulating the partition function, in particular the technique yields the result in Section 4.3. There are however *quicker* and less self-contained approaches. Here we discuss:

(1) The approach by using concentration inequalities, which, properly speaking, is rather a way to prove self-averaging (and it is in any case based on Proposition 4.2). Relatively speaking, it requires restrictive conditions on ω, but it yields very strong probability estimates on the self-averaging phenomenon.
(2) The approach by using Kingman Super-Additive Ergodic Theorem.

We also discuss the validity of the arguments in Section 4.3 beyond the IID set-up.

4.4.1　*Concentration inequalities and self-averaging*

In recent years several concentration inequalities for product measures (and beyond) have been proven under a variety of assumptions. Some of these concentration inequalities are presented in Appendix A.3. An important application of these inequalities, and actually one of the motivations that has driven this field of research, is to prove self-averaging of statistical mechanics quantities, notably the free energy of disordered systems. The proof of the existence of the free energy is therefore reduced to the existence of the quenched averaged free energy (Proposition 4.2).

Concentration inequalities require assumptions on ω and in this sense they yield weaker results. However, they give explicit bounds on the probability of deviation from the typical behavior and this is very important both on a conceptual level and for applications.

Let us for example apply (A.16) to the (disordered) pinning model in the case $\vartheta(t) = c_1 \exp(-c_2 t^2)$. We set $G(\omega) := (1/N) \log Z_{N,\omega}^a$. We need to estimate $|G(\omega) - G(\omega')|$ and we introduce $\omega_t := t\omega + (1-t)\omega'$ for $t \in [0,1]$. By using $G(\omega) - G(\omega') = \int_0^1 (\mathrm{d}G(\omega_t)/\mathrm{d}t)\,\mathrm{d}t$ we obtain

$$
\begin{aligned}
|G(\omega) - G(\omega')| &= \left| \int_0^1 \frac{\beta}{N} \sum_{n=1}^{N} (\omega_n - \omega_n') \, \mathbf{E}_{N,\omega_t}^a[\delta_n] \, \mathrm{d}t \right| \\
&\leq \frac{|\beta|}{N} \sqrt{ \sup_t \sum_{n=1}^{N} \left(\mathbf{E}_{N,\omega_t}^a[\delta_n] \right)^2 \sum_{n=1}^{N} (\omega_n - \omega_n')^2 } \qquad (4.32) \\
&\leq \frac{|\beta|}{\sqrt{N}} |\omega - \omega'|,
\end{aligned}
$$

namely the Lipschitz constant of $G\sqrt{N}/|\beta|$ is (at most) one, which yields

$$
\mathbb{P}\left(\left| \frac{1}{N} \log Z_{N,\omega}^a - \frac{1}{N} \mathbb{E} \log Z_{N,\omega}^a \right| \geq u \right) \leq c_1 \exp\left(-c_2 \frac{u^2 N}{\beta^2} \right), \qquad (4.33)
$$

for every u. Note that (4.33) and Proposition 4.2 imply that $\{(1/N) \log Z_{N,\omega}^a\}_N$ converges in L^p, for every $p \geq 1$ and, by the the first of the Borel–Cantelli Lemmas, also $\mathbb{P}(\mathrm{d}\omega)$-a.s.. But we stress once again the relevance of (4.33) as a finite volume explicit estimate.

4.4.2 *The Super-Additive Ergodic Theorem approach*

Let us apply Theorem A.13 to our context: below (1), (2) and (3) refer to the three assumptions of the theorem. Choose $F_{j,k}(\omega) := \log Z_{j,k,\omega}^c$ and the super-additivity property (2) is an immediate consequence of (4.10). Hypothesis (1) is just the fact that $\omega \sim \theta\omega$ and hypothesis (3) is (4.8). In order to obtain the statement of Theorem 4.1 let us call the F_ω the limit. Since $\mathrm{F}_\omega = \mathrm{F}_{\theta^n \omega}$, then F_ω is measurable with respect to the σ-

algebra $\sigma(\omega_n, \omega_{n+1}, \ldots)$. Since n is arbitrary, F_ω is tail measurable and by Kolmogorov 0-1 law it is (a.s.) a constant and this is another proof of Theorem 4.1.

However with Theorem A.13 we can go much beyond.

- Hypothesis (1) holds if ω is a stationary sequence.
- Hypothesis (2), that is (4.10), holds in great generality: in particular $K(\cdot) > 0$ is largely sufficient (the proof in Section 4.2, while working well beyond the assumptions in Definition 1.4, does require conditions).
- Hypothesis (3), that is (4.8), holds as long as $\sup_N \sum_{n=1}^{N} \mathbb{E}|\omega_n|/N < \infty$.

Therefore for the existence of the free energy F_ω it is sufficient that $\theta\omega \sim \omega$ and that $\omega_1 \in L^1$. Moreover, a sufficient condition for self-averaging is ergodicity, namely that if A is a measurable subset of Ω such that $\theta A \subset A$ then $\mathbb{P}(A) = 0$ or 1. In fact this directly implies that any measurable function on Ω that is invariant under translations is almost surely a constant.

Summing everything up:

Theorem 4.6 *If ω is a stationary ergodic sequence such that $\omega_1 \in L^1$, then the limit of the sequence $\{(1/N) \log Z^c_{N,\omega}\}_N$ exists $\mathbb{P}(\,d\omega)$–a.s. and in L^1. Moreover the limit is non-random.*

Remark 4.7 *The proof in Section 4.2 can be generalized (well) beyond the IID case. Reconsidering the proof one realizes that the key point is that a sufficient condition is that if F is a measurable and integrable function on $\mathbb{R}^M$, then $\lim_n \sum_{j=1}^{n} F(\theta^{jM}(\omega_1, \ldots, \omega_M))/n = \mathbb{E}[F(\omega)]$. And we want this to hold for every M. This holds for totally ergodic sequences, that is for sequences that are ergodic under the action of θ^k, for every k. This condition is clearly stronger than ergodicity, but it holds under suitably mixing conditions, and therefore well beyond independence.*

4.5 Bibliographic Complements

Complements to Section 4.2

The general scheme of proof given in this section to the existence and self-averaging of the free energy has been used at repeated instances in the statistical mechanics literature. For the literature on random polymers, beyond the class of models we treat, we mention [Orlandini *et al.* (2000)] and [Orlandini *et al.* (2002)].

Complements to Section 4.3

It is natural to conjecture that the phenomenon set forth with Figure 4.1 does not happen and that the critical point converges, in the limit $T \to \infty$, to the critical point of the quenched model, but this is an open problem.

Complements to Section 4.4

The theory of concentration of measure has by now a fundamental tool in the analysis of disordered systems, see in particular [Talagrand (2003)] and [Bovier (2006)], going well beyond proving the self-averaging property of the free energy (for further applications in the polymer context see Chapter 8). A proof of the existence of the free energy for a general disordered pinning model via super-additivity and concentration can be found in [Alexander and Sidoravicius (2006)]. See Appendix A.3 for further technical references on concentration. Kingman's Super-Additive (or Sub-Additive) Ergodic Theorem has been has been repeatedly used in the statistical mechanics context (*e.g.* [Bolthausen and den Hollander (1997)]). As pointed out in [Janse van Rensburg *et al.* (2001)], by exploiting Remark 4.7 one can explicitly exhibit deterministic sequences ω for which one obtains the quenched free energy of the disordered model with charges ω that are typical realization of an IID sequence such that (for example) $\mathbb{P}(\omega_1 = +1) = \mathbb{P}(\omega_1 = -1) = 1/2$.

Chapter 5

Disordered Pinning Models: The Phase Diagram

This chapter will be devoted to the analysis of the phase diagram of the disordered pinning model: $\mathbf{P}^a_{N,\omega,\beta,h}$

$$\frac{d\mathbf{P}^a_{N,\omega,\beta,h}}{d\mathbf{P}}(\tau) = \frac{1}{Z^a_{N,\omega,\beta,h}} \exp\left(\sum_{n\in\tau,n\leq N} (\beta\omega_n - h)\right) \mathbf{1}_{\Omega^a_S(N)}, \qquad (5.1)$$

and $a = \mathsf{c}$ or f as usual. The disorder ω satisfies the standard assumptions (Definition 1.17) and we set $\mathrm{C_M} := \sup D_\mathrm{M}$, that is $\mathrm{C_M} = \sup\{t : \mathrm{M}(t) < \infty\}$. Of course in this chapter $\mathrm{M}(\cdot) := \mathrm{M}(\cdot;\omega_1)$. Without loss of generality we assume that $\beta \geq 0$. Recall that the free energy of such a model is denoted by $\mathrm{F}(\beta,h)$ and, as we have already seen, it is natural to set $h_c(\beta) := \sup\{h : \mathrm{F}(\beta,h) > 0\}$, since, by monotonicity of $\mathrm{F}(\beta,\cdot)$, $\mathcal{L} = \{(\beta,h) : h < h_c(\beta)\}$ and $\mathcal{D} = \{(\beta,h) : h \geq h_c(\beta)\}$, cf. Definition 1.18. The case $\beta = 0$ is just the homogeneous pinning model, in particular $\mathrm{F}(0,h) = \mathrm{F}(-h)$ and $h_c(0) = \log \Sigma_K$.

5.1 Preliminary Observations on the Free Energy

We collect in the next proposition a number of *soft* results on $\mathrm{F}(\cdot,\cdot)$ and $h_c(\cdot)$, which is a function from $[0,\infty)$ to $\overline{\mathbb{R}}$. Note in particular that formula (5.3) does not depend on α, the decay exponent of $K(\cdot)$.

Proposition 5.1 *For every β and h we have the bounds*

$$\mathrm{F}(0,h) \leq \mathrm{F}(\beta,h) \leq \mathrm{F}(0, h - \log \mathrm{M}(\beta)). \qquad (5.2)$$

101

Moreover $h_c(\cdot)$ is a convex function and

$$h_c(0) \le h_c(\beta) \le h_c(0) + \log M(\beta). \qquad (5.3)$$

In particular, $h_c(\cdot) \in C^0[0, c_M]$ and $h'_c(0) = 0$.

Proof. First of all note that (5.2) implies (5.3). Moreover the fact that $F(\cdot, \cdot)$ is convex implies that $\mathcal{D} = \{(\beta, h) : F(\beta, h) \le 0\}$ is a convex set. Since clearly $(0, h) \in \mathcal{D}$ for every $h \ge 0$, this proves that $h_c(\cdot)$ is convex and the continuity property follows.

Let us establish the lower bound in (5.2). Remark that the function $\beta \mapsto \mathbb{E} \log Z^f_{N,\omega,\beta,h}$ is convex and that

$$\frac{d}{d\beta} \mathbb{E} \log Z^f_{N,\omega,\beta,h}\Big|_{\beta=0} = \sum_{n=1}^{N} \mathbb{E}\left[\omega_n \mathbf{E}^f_{N,\omega,0,h}[\delta_n]\right] = 0, \qquad (5.4)$$

so $\beta \mapsto \mathbb{E} \log Z^f_{N,\omega,\beta,h}$ is non-decreasing. Therefore also $\beta \mapsto F(\beta, h)$ is non-decreasing and the lower bound is proven.

For what concerns the upper bound, we just apply the annealed bound procedure, as in (1.72). Therefore we have

$$\frac{1}{N} \log \mathbb{E} \log Z^f_{N,\omega} \le \frac{1}{N} \log \mathbb{E}\left[\exp\left(\widetilde{\beta} \sum_{n=1}^{N} \delta_n\right)\right], \qquad (5.5)$$

with $\widetilde{\beta} := \log M(\beta) - h$ and the upper bound follows by considering the asymptotic behavior of both sides in (5.5), as $N \to \infty$. $\qquad \square$

5.2   An Improved Lower Bound on the Free Energy

The result we want to prove is that a quenched sequence of charges, even if they are centered and so on the average the charge is zero, plays successfully in favor of localization (the analogous result for weakly inhomogeneous models is proven in Proposition 3.8). This is of course due to the fact that the typical polymer trajectories target positive charges. We introduce the random variable

$$\widehat{\omega}_n(\beta, h) := \log\left(\frac{K(1)^2 \exp(\beta\omega_n - h) + K(2)}{K(1)^2 + K(2)}\right), \qquad (5.6)$$

and of course $\{\widehat{\omega}_n(\beta, h)\}_n$ is an IID sequence.

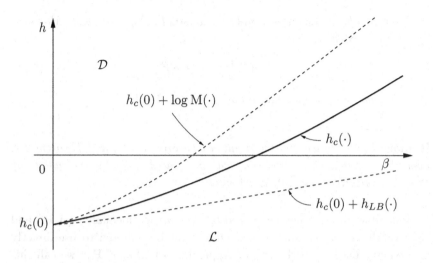

Fig. 5.1 The critical curve with the upper bound in (5.3) and the lower bound of Theorem 5.2. We are using the notation of $h_{LB}(\beta)$ for the supremum of the values of h that satisfy the inequality (5.8).

Clearly if $h < 0$ then $\mathbb{E}\left(\widehat{\omega}_1(\beta, h)\right) > 0$. But the same is true also if $h \geq$ is positive, since ω_1 is non-degenerate (it has variance one), provided that h is sufficiently small and $\beta > 0$. Moreover

$$\mathbb{E}\left(\widehat{\omega}_1(\beta, 0)\right) \overset{\beta \searrow 0}{\sim} \frac{K(1)^2 K(2)}{2(K(1)^2 + K(2))^2} \beta^2. \qquad (5.7)$$

Theorem 5.2

(1) If

$$\mathbb{E}\left(\widehat{\omega}_1(\beta, h)\right) > h \, \frac{2 - K(1)^2}{K(1)^2 + K(2)}, \qquad (5.8)$$

then $(\beta, h + h_c(0)) \in \mathcal{L}$. In particular this implies that

$$h_c(\beta) > h_c(0), \qquad (5.9)$$

for every $\beta > 0$.

(2) For every $\alpha \in (0,1)$ there exists a slowly varying function $\hat{L}(\cdot)$ such that

$$\mathrm{F}(\beta, h_c(0)) \geq \hat{L}(1/\beta)\beta^{2/\alpha}. \qquad (5.10)$$

$\hat{L}(\cdot)$ *may be chosen to be equal to a constant if* $L(\cdot)$ *is trivial. Moreover if* $m_K < \infty$ *then*

$$\mathrm{F}(\beta, h_c(0)) \geq c\beta^2, \qquad (5.11)$$

for some positive constant c *and for every* $\beta \leq 1$.

Remark 5.3 *Notice that the results in Theorem 5.2(2) and Theorem 2.1, applied to the annealed model (i.e. with* β *replaced by* $\log \mathrm{M}(\beta)$*), match, up to a multiplicative constant, for* β *small.*

We follow the strategy suggested by the entropy inequality (A.10) and the variational expression (A.12). One might be tempted to use directly as auxiliary measure ν the one of Appendix B.1, that is $\mathbf{P}_b$: we call this strategy *homogeneous localization*. It is rather easy to see that this strategy cannot work directly: the term $\int X \, \mathrm{d}\nu$ in (A.10) is just the energy of the polymer averaged over the disorder and the contribution of the charges disappear, since they are centered.

The key idea for proving Theorem 5.2 is the following lemma, in which we rewrite $Z^{\mathrm{c}}_{N,\omega}$ as $\mathbf{P}$-expectation of a Boltzmann factor in which the energy contains charges that may have positive $\mathbb{P}$-expectation. Once that is done, the homogeneous localization procedure will yield the desired estimate.

We need the following notation: given τ, for any n odd natural number we set $\hat{\delta}_n := \mathbf{1}_{\{n-1,n+1\}\in\tau}$ and $\hat{\delta}_n := 0$ for n even.

Lemma 5.4 *For* $N \in 2\mathbb{N}$ *the following equality holds*

$$Z^{\mathrm{f}}_{N,\omega,\beta,h} = \mathbf{E}\left[\exp\left(\sum_{n=1}^{N}(\beta\omega_n - h)\delta_n\left(1 - \hat{\delta}_n\right) + \sum_{n=1}^{N}\widehat{\omega}_n(\beta,h)\hat{\delta}_n\right)\right].$$
$$(5.12)$$

Proof. For $N \in 2\mathbb{N}$ and given τ, call $\mathcal{I}_N$ the set of odd sites in $n \in \{1,\ldots,N-1\}$ such that $n-1$ and $n+1$ are in τ. Consider the set that contains 0, N and $n \pm 1$, for every $n \in \mathcal{I}_N$ and number (possibly with repetitions) the points that it contains as $0 =: n_1 \leq n_2 < \ldots < n_{k-1} \leq n_k = N$ with $k = 2|\mathcal{I}_N| + 2$. Note that $n_{2j-1} \leq n_{2j}$, but $n_{2j+1} = n_{2j} + 2$, and that $\mathcal{I}_N = \{n_{2j} + 1 : j = 1,\ldots,|\mathcal{I}_N|\}$. We partition the space of τ–trajectories according to the different values taken up by the (set-valued)

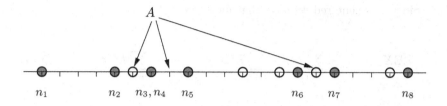

Fig. 5.2 A renewal trajectory up to $N = 20$. The 12 renewal points are identified by circles (some empty, some filled). The 7 filled ones correspond to even renewals that are next nearest neighbors, plus the origin and the end of the *chain* $N = 20$. We have named them $n_1, \ldots, n_8$, with a repetition ($n_1 = 0$, $n_2 = 4$, $n_3 = n_4 = 6$, $n_5 = 8$, $n_6 = 14$, $n_7 = 16$ and $n_8 = 20$). The three sites in between filled circles make the set $A = \{5, 7, 15\}$. The sites in A may or may not host a renewal point. We decompose the space of renewal trajectories according to A: note that, once A is given, between n_1 and n_2, as well as between n_5 and n_6 and between n_7 and n_8 there may be other renewal, but not the ones that modify the set A. For example there cannot be any renewal on the site 18.

random variable $\mathcal{I}_N$, see Figure 5.2. For $A \subset \{1, 3, \ldots, N-3, N-1\}$ we have

$$Z_{N,\omega}^{\mathtt{f}} (\mathcal{I}_N = A) = Z_{n_{2|A|+1}, N, \omega}^{\mathtt{f}, \star} \times$$

$$\left(\prod_{j=1}^{|A|} Z_{n_{2j-1}, n_{2j}, \omega}^{\mathtt{c}, \star} \left(K(1)^2 \exp(\beta \omega_{n_{2j}+1} - h) + K(2) \right) \exp \left(\beta \omega_{n_{2j}+1} - h \right) \right),$$

(5.13)

with $Z_{m,n,\omega}^{a,\star} := Z_{m,n,\omega}^a \left(\sum_{k=m+1}^{n-1} \hat{\delta}_k = 0 \right)$ for $m < n-1$ and $Z_{m,m,\omega}^{a,\star} := 1$. Note that in (5.13) $Z_{m,n,\omega}^{a,\star}$ appears only with m and n that are both even. Observe also that $K(1)^2 \exp(\beta \omega_{n_{2j}+1} - h) + K(2) = (K(1)^2 + K(2)) \exp(\widehat{\omega}_{n_{2j}+1}(\beta, h))$. With this substitution and summing over A one obtains (5.12). $\qquad \square$

Proof of Theorem 5.2. We assume without loss of generality that $\Sigma_K = 1$ (Remark 1.19) and $h \geq 0$. Fix ω and apply (A.10) with $X = X_\omega(\tau)$ equal to the expression in the exponent in the right-hand side of (5.12). The measure ν is the law of τ when the inter-arrival distribution is $K_b(\cdot)$, $b \geq 0$. We are therefore using the strategy of homogeneous localization (Appendix B.1): if $m_K < \infty$ there is no need of localizing and one can take $b = 0$.

Since ω is centered we have that for every τ

$$\mathbb{E}X_\omega(\tau) = -\sum_{n=1}^{N} h\delta_n + \sum_{n=1}^{N} \mathbb{E}\left(\widehat{\omega}_n(\beta, h)\right)\hat{\delta}_n + \sum_{n=1}^{N} h\hat{\delta}_n\delta_n. \tag{5.14}$$

As we have pointed out, $\mathbb{E}\left(\widehat{\omega}_n(\beta, h)\right)$ is strictly positive for certain values of h, including an interval in the positive semi-axis. We are going to average the expression in (5.14) with respect to a positive recurrent renewal: it is then intuitively clear that for h sufficiently small the resulting average will turn out to be positive.

But let us make this quantitative by starting with the $m_K < \infty$ case. Let us observe that

$$\lim_{N\to\infty} \frac{1}{N} \sum_{n=1}^{N} \mathbb{E}\left[\hat{\delta}_n\right] = \frac{K(1)^2 + K(2)}{2m_K}, \tag{5.15}$$

that follows by observing that $\mathbf{P}(\{2n-2, 2n\} \in \tau) = \mathbf{P}(2n-2 \in \tau)(K(1)^2 + K(2))$ and $\mathbf{P}(2n - 2 \in \tau) \stackrel{n\to\infty}{\sim} 1/m_K$ (Theorem A.3). Very much in the same way:

$$\lim_{N\to\infty} \frac{1}{N} \sum_{n=1}^{N} h\mathbb{E}\left[\hat{\delta}_n\delta_n\right] = \frac{K(1)^2}{2m_K}. \tag{5.16}$$

We now choose $b = 0$, that is ν equal to $\mathbf{P}$ in (A.10), so there is no entropic contribution, and we simply have

$$\frac{1}{N}\mathbb{E}\log Z_{N,\omega,\beta,h}^{\mathfrak{f}} \geq \frac{1}{N}\mathbb{E}\mathbf{E}\left[X_\omega(\tau)\right]$$
$$\stackrel{N\to\infty}{=} -\frac{h}{m_K} + \mathbb{E}\left(\widehat{\omega}_1(\beta, h)\right)\frac{K(1)^2 + K(2)}{2m_K} + h\frac{K(1)^2}{2m_K}. \tag{5.17}$$

Therefore, when $m_K < \infty$, $(\beta, h) \in \mathcal{L}$ if the condition (5.8) is satisfied. This proves the first part of Theorem (5.2) when $m_K < \infty$.

For $\alpha \in (0, 1]$ and $m_K = \infty$ we apply the full homogeneous localization strategy, that is $b > 0$. By applying the entropy inequality (A.10) and

Theorem B.1 we obtain

$$
\begin{aligned}
\mathrm{F}(\beta,h) &= \lim_{N\to\infty} \frac{1}{N}\mathbb{E}\log Z^{\mathrm{f}}_{N,\omega,\beta,h} \\
&\geq \lim_{N\to\infty} \frac{1}{N}\mathbb{E}\mathbb{E}_b\left[X_\omega(\tau)\right] - \mathrm{s}(b) \\
&= \frac{1}{m_{K_b}}\left[\mathbb{E}\left(\widehat{\omega}_1(\beta,h)\right)\frac{K_b(1)^2 + K_b(2)}{2} + h\left(\frac{K_b(1)^2}{2} - 1\right)\right] - \mathrm{s}(b).
\end{aligned}
$$
(5.18)

If $\alpha > 0$ the conclusion is immediate: the entropy term is $O(b)$ (Proposition B.2), while m_{K_b} is $O(b^{-1+\alpha})$, except for a slowly varying correction, see (B.14) and the lines following that formula for the case $\alpha = 1$. So as long as the quantity between square brackets in the last line of (5.18) is positive, by choosing b sufficiently close to zero we obtain $\mathrm{F}(\beta,h) > 0$. But since $K_b(n)$ converges to $K(n)$ for $b \searrow 0$, the positivity condition of the quantity between square brackets is implied by (5.8), for b sufficiently small and the proof of the first part of Theorem 5.2 is complete in the case $\alpha > 0$.

For the case $\alpha = 0$, the difficulty comes from the fact that the entropy could be (and in fact is) much larger than b, as suggested by extrapolating (B.9) to $\alpha = 0$. In this case, by using (B.15) and the observation made right after that, we see that the relevant term in the entropy density (B.2), when b is small, is $\Theta(b)\Theta'(b)^{-1}\log\Theta(b)$, which is asymptotically equivalent to $\Theta'(b)^{-1}\log\Theta(b)$. But since $\Theta'(b)^{-1}$ is present both in the energy and in the (relevant part of the) entropy term, it suffices to invoke (B.15) to see that $\log\Theta(b) = o(1)$ and the energy term dominates. So the case $\alpha = 0$ is under control too. This completes the proof of Theorem 5.2 (1).

The proof of Theorem 5.2(2) of course uses (5.7). Formula (5.14) reduces for $h = 0$ to

$$
\mathbb{E}X_\omega(\tau) = \sum_{n=1}^{N} \mathbb{E}\left(\widehat{\omega}_n(\beta,0)\right)\hat{\delta}_n.
$$
(5.19)

Once again the case $m_K < \infty$ is easier, since there is no need of making a change of measure. One readily sees that (5.11) holds if one chooses

$$
c < \frac{1}{2}\left(\frac{K(1)^2 + K(2)}{m_K}\right)\left(\frac{K(1)^2 K(2)}{2(K(1)^2 + K(2))^2}\right),
$$
(5.20)

and β sufficiently small.

If instead $\alpha \in (0,1)$ we restart from (5.18): in the $h = 0$ case the last term behaves, as $b \searrow 0$, like

$$c_1 \frac{b^{1-\alpha}}{L(1/b)} \mathbb{E}\left(\widehat{\omega}_1(\beta, 0)\right) - c_2 b, \qquad (5.21)$$

where c_1 and c_2 are positive constants (depending on α). We have of course applied (B.14) and Proposition B.2. By (5.7), for $\beta < 1$ the term in (5.21) is bounded below by

$$c_3 \frac{b^{1-\alpha}}{L(1/b)} \beta^2 - c_2 b, \qquad (5.22)$$

for a suitable positive constant c_3. If $L(\cdot)$ is trivial, the quantity in (5.22) is bounded below by $c_4 b^{1-\alpha} \beta^2 - c_2 b$, with a suitable choice of c_4. It is now just a matter of choosing the free parameter b such that $\beta^2 = q b^\alpha$, for $q > 0$, obtaining that for $q \in (0, q_0)$ and b (and β) sufficiently small (q_0 may be chosen arbitrarily large, as long as one restricts b to a sufficiently small neighborhood of 0)

$$\mathrm{F}(\beta, 0) \geq (c_4 q - c_2)\, b = (c_4 q - c_2)\, q^{-1/\alpha}\, \beta^{2/\alpha}. \qquad (5.23)$$

By choosing for example $q = 2c_2/c_4$, (5.10) is proven in the case in which $L(\cdot)$ is trivial.

The argument is essentially the same for general $L(\cdot)$, but we should discuss the problem of inverting, *i.e.* solving for b, the equation $\beta^2 = q b^\alpha L(1/b)$. For this we apply L.5 of the list the basic properties of slowly varying functions (Appendix A.4). Therefore we know that $b \mapsto b^\alpha L(1/b)$ is strictly decreasing, and therefore invertible, for b sufficiently small. The inverse map is $x \mapsto x^{1/\alpha} \hat{L}(1/x)$, defined for $x > 0$ in a neighborhood of 0, with $\hat{L}(\cdot)$ a slowly varying function. Therefore $b = \beta^{2/\alpha} q^{-1/\alpha} \hat{L}(q/\beta^2)$, and since $q > 0$ is fixed, $\hat{L}(q/\beta^2) \sim \hat{L}(1/\beta^2)$ as $\beta \searrow 0$. With this change of variables the expression in (5.22) transforms into $(c_3 q - c_1)b$. The choice of $q = 2c_1/c_3$ leads, redefining $\hat{L}(\cdot)$, to (5.10) and the proof of Theorem 5.2 is complete.

Theorem 5.2

$\square$

5.3 On the Upper Bound: Annealing and Constrained Annealing

A procedure that can improve the annealed estimate (5.5) is the so called *constrained annealing*. Here we discuss a form of constraining based on introducing suitable *Lagrange multipliers*. The idea is based on realizing that the limit of the annealing procedure, discussed in Section 1.7.1, comes from the fact that in reality the ω configurations that contribute to the annealed estimates are atypical ones. Let us make this explicit for completeness: for $\Sigma_K = 1$, $h < \log M(\beta)$ and β in the interior of D_M (see Definition 1.16)

$$
\lim_{N \to \infty} \mathbb{E}\mathbf{E}_{N,\omega}^{\mathrm{f}} \left[\frac{1}{N} \sum_{n=1}^{N} \omega_n \right]
$$

$$
= \lim_{N \to \infty} \frac{\partial}{\partial \lambda} \frac{1}{N} \log \mathbb{E}\mathbf{E} \left[\exp \left(\sum_{n=1}^{N} (\beta \omega_n - h) \delta_n + \lambda \sum_{n=1}^{N} \omega_n \right) \right] \Big|_{\lambda=0}
$$

$$
= \lim_{N \to \infty} \frac{\partial}{\partial \lambda} \frac{1}{N} \log \mathbf{E} \left[\exp \left(\sum_{n=1}^{N} (\log M(\beta \delta_n + \lambda) - h \delta_n) \right) \right] \Big|_{\lambda=0}
$$

$$
= \frac{M'(\beta)}{M(\beta)} \lim_{N \to \infty} \mathbf{E}_{N,\log M(\beta) - h}^{\mathrm{f}} \left[\frac{1}{N} \sum_{n=1}^{N} \delta_n \right] > 0. \quad (5.24)
$$

So under the annealed measure the charges are no longer centered and this shows that the annealing procedure has a rather drastic effect on the disorder. In order to try to prevent this and, possibly, improve the upper bound on the free energy, one typically introduces some constraint. We can for example implement this idea by inserting some terms that can be interpreted as Lagrange multipliers. Following this idea, let us start by observing that given a sequence of centered random variables $\{A_N(\omega)\}_N$ we have

$$
\mathrm{F}(\beta, h) \leq \liminf_{N \to \infty} \frac{1}{N} \log \mathbb{E} \left[Z_{N,\omega}^a \exp \left(A_N(\omega) \right) \right]. \quad (5.25)
$$

We have simply used Jensen inequality, the convergence in L^1 of $\{(1/N) \log Z_{N,\omega}^a\}_N$ and the fact that $\mathbb{E} \left[\log \left(Z_{N,\omega}^a \exp \left(A_N(\omega) \right) \right) \right]$ is equal to $\mathbb{E} \log Z_{N,\omega}^a$. In words: modifying the Hamiltonian by adding the disorder dependent term $A_N(\omega)$ does not modify the quenched averaged free energy (which therefore coincides with the quenched free energy of the original model). However the bound (5.25) can improve on the bound we find with $A_N \equiv 0$. This can be seen immediately by making the important observation that such a bound can be saturated: it suffices to choose $A_N(\omega) = -\log Z_{N,\omega}^a + \mathbb{E} \log Z_{N,\omega}^a$.

In practice, in order to take profit of (5.25) we have to deal with *amenable* random variables $A_N(\omega)$. A reasonable (and often proposed) choice is the following: take $G : \Omega \to \mathbb{R}$ such that $G(\omega)$ is a centered random variable and set $A_N(\omega) := \sum_{n=1}^{N} G(\theta^n \omega)$. In order for this class of random variables to be worth considering we have to add at least the condition that $G(\cdot)$ is local, *i.e.* that there exists $n \in \mathbb{N}$ such that $G(x) = G(y)$ whenever x and $y \in \mathbb{R}^{\mathbb{N}}$ satisfy $x_i = y_i$ for $i = 1, \ldots, n$. In reality, computations are in general extremely difficult beyond the easiest example $G(x) = const.x_1$.

Let us show with an example that a good (actually, the most elementary) choice of $G(\cdot)$ can improve quantitatively on the upper bound on $\mathrm{F}(\cdot)$ that we find with $G(\cdot) \equiv 0$. Choose $G(\omega) = \lambda \omega_1$, $\lambda \in \mathbb{R}$. By direct computation we see that

$$\mathbb{E}\left[Z_{N,\omega}^a \exp\left(\sum_{n=1}^{N} G(\theta^n \omega)\right)\right] = \mathrm{M}(\lambda)^N \mathbb{E}\left[\exp\left(\sum_{n=1}^{N} \widetilde{\beta}\delta_n\right)\right], \quad (5.26)$$

with $\widetilde{\beta} = \log \mathrm{M}(\beta + \lambda) - \log \mathrm{M}(\lambda) - h$, so that (5.25) yields

$$\mathrm{F}(\beta, h) \leq \inf_{\lambda \in \mathbb{R}} \left[\log \mathrm{M}(\lambda) + \mathrm{F}\left(\log \mathrm{M}(\beta + \lambda) - \log \mathrm{M}(\lambda) - h\right)\right]. \quad (5.27)$$

Set $h = 0$, consider small values of β (recall that $\mathrm{var}(\omega_1) = 1$) and choose $K(\cdot)$ such that $m_K < \infty$, so that $\mathrm{F}(\delta) \overset{\delta \searrow 0}{\sim} \delta/m_K$ (Theorem 2.1). It is then evident that

$$\limsup_{\beta \searrow 0} \frac{1}{\beta^2} \mathrm{F}(\beta, 0) \leq \frac{1}{2m_K}\left(1 - \frac{1}{m_K}\right), \quad (5.28)$$

while annealing with $\lambda = 0$ yields the bound $1/(2m_K)$ on the same quantity. So the procedure is not void of significance.

While the observations we have just made may leave hopes of strong improvements, the following result goes in the opposite direction. Recall that $\lim_{N \to \infty}(1/N) \log \mathbb{E} Z_{N,\omega}^a = \mathrm{F}(\log \mathrm{M}(\beta) - h)$.

Proposition 5.5 *If* $\mathrm{F}(\log \mathrm{M}(\beta) - h) > 0$ *then for every local bounded function* $G(\cdot)$, *with* $G(\omega)$ *centered random variable, we have*

$$\lim_{N \to \infty} \frac{1}{N} \log \mathbb{E}\left[Z_{N,\omega}^a \exp\left(\sum_{n=1}^{N} G(\theta^n \omega)\right)\right] > 0. \quad (5.29)$$

In short, the upper bound on the free energy may indeed get better by constrained annealing, but this procedure does not lead to a better estimate on the location of the critical point (or line). This leaves open the possibility of obtaining $F(\beta, h) = 0$ by taking the infimum over $G(\cdot)$ in the left-hand side of (5.29). But the reader may easily realize how difficult is going beyond the example we have worked out, so the possibility we are setting forth has only a theoretical value. We stress also that having chosen $G(\cdot)$ bounded makes the statement particularly simple: one can in principle replace it with suitable integrability conditions. And there is no loss of generality in choosing $G(\cdot)$ bounded in the case in which ω_1 takes only a finite number of values.

We will not give a full proof of this statement, but the idea is very simple: it is based on

$$\mathbb{E}\left[Z_{N,\omega}^{\mathrm{f}} \exp\left(A_N(\omega)\right)\right] \geq \mathbf{P}(\tau_1 > N)\mathbb{E}\left[\exp\left(A_N(\omega)\right)\right] \overset{N\to\infty}{\asymp} \mathbb{E}\left[\exp\left(A_N(\omega)\right)\right].$$
(5.30)

Now set $A_N(\omega) = \sum_{n=1}^{N} G(\theta^n \omega)$. Then it is easy to see that the limit of $\{(1/N)\log\mathbb{E}\left[\exp\left(A_N(\omega)\right)\right]\}_N$ exists and (by Jensen inequality) it is non-negative. If it is positive then the statement is proven. If instead it is equal to zero then one can show that there exists a local bounded function $Q : \Omega \to \mathbb{R}$ such that $G(x) = Q(\theta x) - Q(x)$, that is $G(\cdot)$ is a gradient and therefore $\sum_{n=1}^{N} G(\theta^n \omega) = O(1)$ and it gives no contribution to the Laplace asymptotic behavior of the left-hand side of (5.30), hence falling back to the case of $A_N \equiv 0$ and the constrained annealing coincides with the annealing. For a bibliography on this, see Section 5.7.

5.4 The Effect of the Disorder on the Order of the Transition: A Smoothing Inequality

In this section we are going to prove the following result:

Theorem 5.6 *If ω_1 is a continuous random variable and if there exists $R > 0$ and $\varepsilon > 0$ such that the relative entropy of the law of $\omega_1 + x$ with respect to the law of ω_1 is bounded by Rx^2 for $|x| \leq \varepsilon$, then for every $\beta > 0$ and every $\alpha \geq 0$*

$$F(\beta, h) \leq (1 + \alpha)R\beta^{-2}\left(h_c(\beta) - h\right)^2,$$
(5.31)

for h such that $0 < h_c(\beta) - h \leq \varepsilon\beta$.

Remark 5.7   *Theorem 5.6 can be proven in greater generality, notably for general bounded disorder variables ω (see Section 5.7). The proof turns out to be less technical under the relative entropy condition in Theorem 5.6. We spell out this condition explicitly here: for $|x| \leq \varepsilon$ we require*

$$\int_{\mathbb{R}} f_{\omega_1}(r+x) \log \left(\frac{f_{\omega_1}(r+x)}{f_{\omega_1}(r)} \right) dr \leq Rx^2. \qquad (5.32)$$

If $f_{\omega_1}(r) = \exp(-V(r))$ for some C^2 function $V(\cdot) : \mathbb{R} \to \mathbb{R}$ then, by Taylor formula, we see that we can choose

$$R = \frac{1}{2} \int_{\mathbb{R}} \exp\left(-V(r)\right) \max_{|x| \leq \varepsilon} |V''(r+x)| \, dr. \qquad (5.33)$$

So $R < \infty$ for very general functions $V(\cdot)$, in particular whenever $V(\cdot)$ is a polynomial function. Note that in the case of $\omega_1 \sim \mathcal{N}(0,1)$ one can choose $\varepsilon = \infty$ and $R = 1/2$.

Before proving Theorem 5.6 let us discuss its relevance. For this it is useful to go back to Proposition 1.6 or to Theorem 2.1: these results, giving the behavior of the free energy of the homogeneous pinning model near the critical point, are summed up in Figure 1.6. If one rewrites (5.31) as

$$0 \leq \mathrm{F}(\beta, h) - \mathrm{F}(\beta, h_c(\beta)) \leq (1+\alpha)R\beta^{-2} \left(h_c(\beta) - h\right)^2, \qquad (5.34)$$

it becomes evident that it is an estimate on the regularity of the free energy at criticality. More precisely it says that, regardless of the value of α, $\mathrm{F}(\beta, \cdot)$ is C^1 with derivative which is Lipschitz continuous at $h_c(\beta)$, so the transition is at least of second order (almost third: the regularity is at least the one of the homogeneous model with $\alpha = 1/2$ and trivial $L(\cdot)$). So the disorder has really a *smoothing* effect if $\alpha > 1/2$, see Figure 5.4.

Proof of Theorem 5.6. Choose $\beta > 0$. The key point is the following Large Deviations estimate of observing an atypically large value of the finite volume free energy:

$$\liminf_{\ell \to \infty} \frac{1}{\ell} \log \mathbb{P} \left(\log Z^c_{\ell,\omega,\beta,h} \geq a\ell\mathrm{F}(\beta, h - \delta) \right) \geq -R \left(\frac{\delta}{\beta} \right)^2, \qquad (5.35)$$

for any $a \in (0,1)$, $\delta \in (0, \beta\varepsilon]$ and h such that $\mathrm{F}(\beta, h - \delta) > 0$. This follows from (A.13) by choosing $\mu(= \mu_\ell)$ equal to the law of $(\omega_1, \dots, \omega_\ell)$, $\nu(= \nu_\ell)$ equal to the law of $(\omega_1 + (\delta/\beta), \dots, \omega_\ell + (\delta/\beta))$ and of course $E = E_\ell := \left\{ \omega : \log Z^c_{\ell,\omega,\beta,h} \geq a\ell\mathrm{F}(\beta, h - \delta) \right\}$. In fact $\nu_\ell(E_\ell)$ tends to 1 as $\ell \to \infty$, which is a consequence of Theorem 4.1, of the choice of a and of $\beta\omega_n -$

$(h - \delta) = \beta(\omega_n + (\delta/\beta)) - h$. Moreover, by hypothesis and independence, we have the bound $\mathcal{S}\,(\nu_\ell|\mu_\ell) \leq \ell R(\delta/\beta)^2$, so that (5.35) is established.

The inequality (5.35) gives a lower bound on how frequently we will observe atypically large values of the finite volume free energy. Namely, let us choose ℓ sufficiently large that

$$p(\ell) := \mathbf{P}^{\mathrm{c}}_{\ell,\omega,\beta,h}\left(\log Z^{\mathrm{c}}_{\ell,\omega,\beta,h} \geq a\ell\mathrm{F}(\beta, h - \delta)\right) \geq \exp\left(-\ell R\delta^2/(a\beta^2)\right),$$
(5.36)

and that

$$\log K\,(j\ell) \geq -\frac{1}{a}(1+\alpha)\log(j\ell), \quad \text{for every } i \in \mathbb{N}.$$
(5.37)

Of course (5.37) is satisfied for ℓ sufficiently large by the elementary property of slowly varying functions spelled out in (1.25). We stress that the parameter a in the rightmost term of (5.36) and in (5.37) is an arbitrary number in $(0,1)$, chosen, for conciseness, equal to the other arbitrary quantity appearing in the middle term in (5.36).

We now introduce the sequence of IID random variables, in fact Bernoulli random variables, $\{Y_j(\omega)\}_{j=1,2,\ldots}$ with $Y_j(\omega) := \mathbf{1}_{\theta(j-1)\ell\omega \in E_\ell}$. We may and will assume that $N/\ell \in \mathbb{N}$: of course we are interested only on $\{Y_j(\omega)\}_{j=1,2,\ldots,N/\ell}$ so, when we refer to $Y_j(\omega)$, j spans from 1 to N/ℓ. Let $B_\ell(\omega)$ be the union of the sets $\{\ell(j-1),\ldots \ell j\}$ with $Y_j = 1$ and consider the set $\Omega_S(\omega, \ell, N)$ of τ trajectories defined by the following requirements (see also Figure 5.3):

(1) $\tau \cap (\{1,\ldots,N-1\} \setminus B_\ell(\omega)) = \emptyset$, that is the polymer visits the defect line only in the ℓ-blocks with $Y_j(\omega) = 1$ (*rich blocks*);
(2) If $Y_j(\omega) = 1$ then $\{(j-1)\ell, j\ell\} \in \tau$, namely the polymer visits the border of each block which is rich (we call *standard* a block which is not rich).

We remark that of course $Z^{\mathrm{c}}_{N,\omega} \geq Z^{\mathrm{c}}_{N,\omega}\,(\Omega_S(\omega, \ell, N))$ (from now on in this proof all partition functions are meant as computed at (β, h)). The latter quantity may be easily estimated, by using the renewal property at the boundaries of the rich blocks, but we still need some notation. Call $G_n(= G_n(\omega))$ the random variable taking values in $\mathbb{N} \cup \{0\}$ that counts the number of standard blocks separating the $(n-1)^{\mathrm{th}}$ and the nth rich block (in this sense we imagine that the origin is contained in the 0th rich block that lies in the negative half-line).

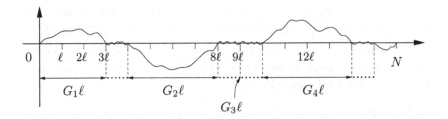

Fig. 5.3   The set of trajectories $\Omega_S(\omega, \ell, N)$. We have drawn a polymer trajectory and not only the τ trajectory. In this case four blocks are rich, they correspond with the (horizontal) dotted lines, and $G_1 = 3$, $G_2 = 4$, $G_3 = 0$ and $G_4 = 4$. The selected trajectories touch the defect line at 3ℓ, 4ℓ, 8ℓ, 9ℓ, 10ℓ, 14ℓ and 15ℓ and they cannot touch the line out of the rich blocks (with exception of course of 0 and $N = 16\ell$). As ℓ becomes large, the rich blocks become rare, so we call this lower bound strategy a *rare stretch strategy*: it will also play a role in Chapter 6.

Remark 5.8   *In order to be precise let us give a formal definition of* $(G_1, \ldots, G_{\mathcal{N}_Y(\omega)})$. *They are defined by means of the auxiliary variables* D: $D_0 := 0$ *and* $D_{n+1} := \inf\{j > D_n : Y_j = 1\}$. *In fact* $G_n = D_n - D_{n-1} - 1$. *Note that* $G_n \leq N/\ell$ *if and only if* $n \leq \mathcal{N}_Y(\omega)$ *and those are the only variables that interest us.*

Therefore we have the bound:

$$\log Z_{N,\omega}^{\mathrm{c}} \geq \sum_{j:\, Y_j(\omega)=1} \log Z_{\ell,\theta^{(j-1)\ell}\omega}^{\mathrm{c}} + \sum_{n=1}^{\mathcal{N}_Y(\omega)} \log\left(K(G_n\ell)\right) + R_N(\omega,\ell),$$

(5.38)

where $K(0)$ has to be read as 1 and $R_N(\omega, \ell)$ is the contribution due to the last excursion. Notice immediately that $R_N(\omega, \ell) \geq -c \log N$, for some $c > 0$, uniformly in ω and ℓ, so this term is totally irrelevant. The other important observation is that on rich blocks we have a lower bound on the partition function, so that we obtain

$$\log Z_{N,\omega}^{\mathrm{c}} \geq a\ell \mathrm{F}(\beta, h - \delta)\mathcal{N}_Y(\omega) + \sum_{n=1}^{\mathcal{N}_Y(\omega)} \log\left(K(G_n\ell)\right) + O(\log N), \quad (5.39)$$

where $\mathcal{N}_Y(\omega)$ is the cardinality of $\{j : Y_j(\omega) = 1\}$. We now observe that, by (5.37) and then by Jensen's inequality, if we set $\hat{G}_n := \max(G_n, 1/\ell)$ we have

$$\sum_{n=1}^{\mathcal{N}_Y(\omega)} \log\left(K(G_n\ell)\right) \geq -\frac{1+\alpha}{a}\mathcal{N}_Y(\omega)\left(\frac{1}{\mathcal{N}_Y(\omega)}\sum_{n=1}^{\mathcal{N}_Y(\omega)}\log\left(\hat{G}_n\ell\right)\right)$$

$$\geq -\frac{1+\alpha}{a}\mathcal{N}_Y(\omega)\log\left(\frac{1}{\mathcal{N}_Y(\omega)}\sum_{n=1}^{\mathcal{N}_Y(\omega)}\hat{G}_n\ell\right) \qquad (5.40)$$

$$\geq -\frac{1+\alpha}{a}\mathcal{N}_Y(\omega)\log\left(\frac{N}{\mathcal{N}_Y(\omega)}\right),$$

where $\hat{G}_n$ has been introduced to match with the definition of $\log K(0) = 0$ used in this proof. Note that in the last step we have used $\sum_{n=1}^{\mathcal{N}_Y(\omega)}G_n\ell = N - \ell\mathcal{N}_Y(\omega)$ and that $\sum_{n=1}^{\mathcal{N}_Y(\omega)}\hat{G}_n\ell$ differs from the previous quantity of less than $\mathcal{N}_Y(\omega)$, so that $\sum_{n=1}^{\mathcal{N}_Y(\omega)}\hat{G}_n\ell \leq N$.

By putting (5.39) and (5.40) together and by using

$$\lim_{N\to\infty}\frac{1}{N}\mathcal{N}_Y(\omega) = \frac{p(\ell)}{\ell}, \qquad (5.41)$$

$\mathbb{P}(\,d\omega)$–a.s., which follows directly from the Strong Law of Large Numbers, we obtain

$$\mathrm{F}(\beta,h) \geq p(\ell)\left(a\mathrm{F}(\beta,h-\delta) - \frac{1+\alpha}{a}\frac{1}{\ell}\log\left(\ell/p(\ell)\right)\right), \qquad (5.42)$$

and by (5.36) we get to

$$\mathrm{F}(\beta,h) \geq p(\ell)\left(a\mathrm{F}(\beta,h-\delta) - R\frac{1+\alpha}{a^2}\left(\frac{\delta}{\beta}\right)^2 - \frac{1+\alpha}{a}\left(\frac{1}{\ell}\log\ell\right)\right). \qquad (5.43)$$

Now set $h = h_c(\beta)$, so that the left-hand side in (5.42) is zero. We therefore have

$$a\mathrm{F}(\beta,h_c(\beta)-\delta) - R\frac{1+\alpha}{a^2}\left(\frac{\delta}{\beta}\right)^2 \leq \frac{1+\alpha}{a}\left(\frac{1}{\ell}\log\ell\right). \qquad (5.44)$$

Since ℓ may be chosen arbitrarily large, the left-hand side in (5.44) is non-positive, which, by letting $a \nearrow 1$, yields (5.31). $\square$

5.5 What is One Expecting: The Renormalization Group Viewpoint and the Harris Criterion

In [Harris (1974)] a heuristic argument has been proposed to understand the role of disorder in local models. Harris' work targets the Ising model with disordered interaction. It has been claimed at several instances, see Section 5.7, that the analog of Harris' argument, often called Harris criterion, for disordered pinning models yields that the disorder is *relevant* if $\alpha > 1/2$ and it is *irrelevant* if $\alpha < 1/2$; at least for β sufficiently small. What one means for relevant and irrelevant is that the *critical exponents* of the models coincide with the ones of the annealed model. According to the physicists' claim, in this case one expects also that if $\alpha < 1/2$ then the quenched critical curve coincides with the annealed one (again: for β sufficiently small).

Without pretending that we are really following closely Harris' steps, we give a heuristic argument in favor of the fact that $\alpha = 1/2$ is the crossover point for the relevance of the disorder, see Figure 5.4. We consider only the case $\omega_1 \sim \mathcal{N}(0,1)$.

The argument is based on the elementary formula

$$
\mathrm{F}(\beta,h) - \mathrm{F}^{\mathrm{a}}_\beta(h) = \lim_{N\to\infty} \frac{1}{N} \mathbb{E} \log \mathbf{E}^{\mathrm{f}}_{N,\beta_{\mathrm{a}}} \left[\exp\left(\sum_{n=1}^{N} \left(\beta\omega_n - \frac{\beta^2}{2} \right) \delta_n \right) \right],
$$
(5.45)

where $\mathrm{F}^{\mathrm{a}}_\beta(h) := \mathrm{F}((\beta^2/2) - h)$ is the annealed free energy. We set also $\beta_{\mathrm{a}} := \beta_{\mathrm{a}}(\beta,h) = (\beta^2/2) - h$. For $j \in \{1,\ldots,N-1\}$ we introduce the measure

$$
\widetilde{\mathbf{P}}_{N,j,\omega}(A) := \frac{\mathbf{E}^{\mathrm{f}}_{N,\beta_{\mathrm{a}}} \left[\exp\left(\sum_{n=1}^{j} \left(\beta\omega_n - \frac{\beta^2}{2} \right) \delta_n \right) ; A \right]}{\mathbf{E}^{\mathrm{f}}_{N,\beta_{\mathrm{a}}} \left[\exp\left(\sum_{n=1}^{j} \left(\beta\omega_n - \frac{\beta^2}{2} \right) \delta_n \right) \right]},
$$
(5.46)

for $A \subset \Omega_S$ and $\widetilde{\mathbf{P}}_{N,0,\omega} := \mathbf{P}^{\mathrm{f}}_{N,\beta_{\mathrm{a}}}$. With this notation we have

$$
\mathrm{F}(\beta,h) - \mathrm{F}^{\mathrm{a}}_\beta(h) = \lim_{N\to\infty} \frac{1}{N} \mathbb{E} \sum_{j=1}^{N} \log \widetilde{\mathbf{E}}_{N,j-1,\omega} \left[\exp\left(\delta_j \left(\beta\omega_j - \frac{\beta^2}{2} \right) \right) \right].
$$
(5.47)

Since $\delta_j \in \{0, 1\}$ we can write

$$\widetilde{\mathbf{E}}_{N,j-1,\omega} \left[\exp \left(\delta_j \left(\beta\omega_j - \frac{\beta^2}{2} \right) \right) \right] = 1 + \left(\frac{\exp (\beta\omega_j)}{\mathbb{E} \exp (\beta\omega_j)} - 1 \right) \mathbf{E}_{N,j-1,\omega} [\delta_j].$$

$$(5.48)$$

Let us call $X_j(\omega)$ the quantity between parentheses in the right-hand side. Note that $\mathbb{E}[X_j(\omega)] = 0$ and that $\mathbb{E}\left[(X_j(\omega))^2\right] = \exp(\beta^2) - 1 \overset{\beta \searrow 0}{\sim} \beta^2$. Moreover $X_j(\omega)$ and $\mathbf{E}_{N,j-1,\omega}[\delta_j]$ are independent so by Taylor expansion we get for β small

$$\mathbb{E} \log \widetilde{\mathbf{E}}_{N,j-1,\omega} \left[\exp \left(\delta_j \left(\beta\omega_j - \frac{\beta^2}{2} \right) \right) \right] =$$
$$- \frac{1}{2} \left(\exp(\beta^2) - 1\right) \mathbb{E}\left[\widetilde{\mathbf{E}}_{N,j-1,\omega}[\delta_j]^2\right] + \text{higher order terms.} \quad (5.49)$$

We make the hypothesis that for small β the higher order terms are negligible and that $\mathbf{E}_{N,j-1,\omega}[\delta_j]$ is close to $\mathbf{E}^{\mathrm{f}}_{N,\beta_{\mathrm{a}}}[\delta_j]$ (we are essentially injecting the idea that the disorder is irrelevant). Note that, in the limit as $N \to \infty$, $\mathbf{E}^{\mathrm{f}}_{N,\beta_{\mathrm{a}}}[\delta_j]$ is the annealed contact fraction:

$$\mathrm{N}^{\mathrm{a}}_\beta(h) := \mathrm{N}((\beta^2/2) - h) = \frac{\mathrm{d}}{\mathrm{d}h} \mathrm{F}^{\mathrm{a}}_\beta(h), \quad (5.50)$$

and we recall that $\mathrm{N}(\cdot)$ is the contact fraction of the homogeneous pinning model. Therefore the steps from (5.45) to (5.49) tell us that if our hypothesis is true then

$$\mathrm{F}(\beta, h) - \mathrm{F}^{\mathrm{a}}_\beta(h) = -\frac{1}{2}\beta^2 \left(\mathrm{N}^{\mathrm{a}}_\beta(h)\right)^2 + \text{higher order terms.} \quad (5.51)$$

Our hypothesis is consistent only if $\mathrm{F}^{\mathrm{a}}_\beta(h) - \mathrm{F}(\beta, h) \ll \mathrm{F}^{\mathrm{a}}_\beta(h)$, namely if

$$\beta^2 \left(\mathrm{N}^{\mathrm{a}}_\beta(h)\right)^2 \ll \mathrm{F}^{\mathrm{a}}_\beta(h). \quad (5.52)$$

Let us put ourselves very close to the annealed critical point $h = \beta^2/2 =: h^{\mathrm{a}}_c(\beta)$ (β is small but fixed). Choosing $L(\cdot)$ (in the definition of $K(\cdot)$) trivial, by Theorem 2.1, we have

$$\mathrm{F}^{\mathrm{a}}_\beta(h^{\mathrm{a}}_c(\beta) - \delta) \overset{\delta \searrow 0}{\sim} const.\ \delta^{\max(1/\alpha, 1)} \quad (5.53)$$

and by Theorem 2.4

$$\mathrm{N}^{\mathrm{a}}_\beta(h^{\mathrm{a}}_c(\beta) - \delta) \overset{\delta \searrow 0}{\sim} const.\ \delta^{\max((1/\alpha) - 1, 0)}. \quad (5.54)$$

Of course (5.52) fails if $\alpha \geq 1$. For $\alpha \in (0,1)$ instead it holds if $\delta^{(2/\alpha)-2} \ll \delta^{1/\alpha}$, that is if $\alpha < 1/2$.

So the conclusion is that $\alpha = 1/2$ should be the crossover point between irrelevant and relevant disorder (see Figure 5.4).

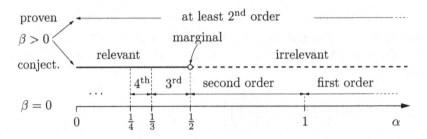

Fig. 5.4   The smoothing effect of quenched disorder. Theorem 5.6 says that the transition, as soon as $\beta > 0$, is of second order or higher, so there is a smoothing effect for $\alpha > 1/2$, in the sense that the transition for $\beta = 0$ is less regular than the transition for $\beta > 0$. It is interesting to note that the values of α for which smoothing is proven (thick dashed line) coincide with the value for which the noise is relevant in the sense of Harris. For $\alpha < 1/2$ the noise is instead irrelevant, in the same sense, and it is conjectured that the critical exponents, and therefore the order of the transition, should not change if a small disorder is switched on.

5.6   A Fully Inhomogeneous Non-Disordered Model

Disordered models are very fashionable and they have the undeniable merit of allowing substantial progress in analyzing inhomogeneous systems. Of course the results one obtains may depend strongly on the choice of the distribution of the disorder and beyond the differences that may arise by changing the law of ω_1 in the IID distributed set-up (that is at the heart of our approach), it is rather easy to see that introducing (strong) correlations in the sequence of charges leads in most of the cases to very different scenarios. But if one can in principle try to analyze large classes of disordered charges, the disordered systems approach carries with itself a fundamental drawback: results are established for *almost every* realization of the disorder, but rarely one can single out a particular realization for which the results hold. Examples of such rare cases can be built by using Remark 4.7, see Section 4.5, but these examples are very particular and somewhat unnatural.

Here instead we give a general result on the following case: consider the general pinning model (5.1). In this section we suppose, for ease of exposition, that $\Sigma_K = 1$, so the underlying free renewal is recurrent. We have seen that $h_c(\beta) \geq 0$ for every β, so the system is localized for $h < 0$. This directly implies that if $\widetilde{\omega} := \{\widetilde{\omega}_n\}_n$ is an IID sequence of Bernoulli variables of parameter $p \in (0, 1]$, that is $\mathbb{P}(\widetilde{\omega}_1 = 1) = p = 1 - \mathbb{P}(\widetilde{\omega}_1 = 0)$, then the polymer with measure $\mathbf{P}^a_{N,\widetilde{\omega},\lambda}$

$$\frac{d\mathbf{P}^a_{N,\omega,\lambda}}{d\mathbf{P}}(\tau) = \frac{1}{Z^a_{N,\omega,\lambda}} \exp\left(\lambda \sum_{n \in \tau, n \leq N} \widetilde{\omega}_n\right) \mathbf{1}_{\Omega^a_S(N)}, \tag{5.55}$$

with $\lambda > 0$, is localized for almost every realization of $\widetilde{\omega}$. Note that this is a diluted pinning model and the result we just stated says that an arbitrarily large disordered dilution cannot lead to delocalization.

However a (much) stronger result holds.

Theorem 5.9 *Choose $K(\cdot)$ as in Definition 1.4 and assume that $\Sigma_K = 1$ and $m_K = \infty$, so that the free process is null recurrent. Moreover assume that $\alpha > 0$. Let $\widetilde{\omega}$ be a numeric sequence taking values in $\{0, 1\}^{\mathbb{N}}$, $\lambda > 0$ and $a = $ c or f. The following three statements are equivalent:*

$$\liminf_{N \to \infty} \frac{1}{N} \sum_{n=1}^{N} \widetilde{\omega}_n > 0, \tag{5.56}$$

$$\liminf_{N \to \infty} \frac{1}{N} \log Z^a_{N,\widetilde{\omega},\lambda} > 0, \tag{5.57}$$

$$\liminf_{N \to \infty} \mathbf{E}^a_{N,\widetilde{\omega},\lambda}\left[\frac{\mathcal{N}_N(\tau)}{N}\right] > 0. \tag{5.58}$$

We therefore have a criterion for localization for general pinning models with arbitrary (deterministic!) dilution. For the case $\alpha = 0$ see Section 5.7. Note moreover that if the underlying renewal is positive recurrent, that is $m_K < \infty$, then the model $\mathbf{E}^a_{N,\widetilde{\omega},\lambda}[\mathcal{N}_N(\tau)] \geq \mathbf{E}^a_{N,\widetilde{\omega},0}[\mathcal{N}_N(\tau)]$ and the latter expression is just the expectation of the number of renewals up to times N for the free process, a quantity which is asymptotically equivalent to N/m_K.

The fundamental estimate is the following:

Lemma 5.10 *Under the hypotheses of Theorem 5.9, if there exists $\delta > 0$ and $N_0 \in \mathbb{N}$ such that*

$$\inf_{N > N_0} \frac{1}{N} \sum_{n=1}^{N} \widetilde{\omega}_n \geq \delta, \tag{5.59}$$

then there exists $\mathrm{c} = \mathrm{c}(\lambda, \delta) > 1$ such that $Z_{N,\widetilde{\omega},\lambda}^a \geq \mathrm{c}^N$ for $N > N_0$.

Proof. For $\widetilde{\omega}$ satisfying (5.59), we set $E_N := \{n \in \{1, \ldots, N\} : \widetilde{\omega}_n = 1\}$. Observe that we can write

$$\exp\left(\lambda \sum_{n \in \tau:\, n \leq N} \widetilde{\omega}_n\right) = \prod_{n=1}^{N} \left((\exp(\lambda) - 1)\, \mathbf{1}_{n \in \tau} \mathbf{1}_{n \in E_N} + 1\right), \tag{5.60}$$

and expanding the product we obtain the useful representation formula:

$$Z_{N,\widetilde{\omega},\lambda}^a = \sum_{A \subset E_N} (\exp(\lambda) - 1)^{|A|} \, \mathbf{P}\left(A \subset \tau,\, \Omega_S^a(N)\right). \tag{5.61}$$

Let us consider the case $a = \mathbf{f}$ (the other case is almost identical). Write $A =: \{n_1, n_2, \ldots, n_{|A|}\}$, with $n_j > n_{j-1}$, and set $n_0 = 0$. Let us also fix $|A| =: n$. By the renewal property we have

$$\mathbf{P}\left(A \subset \tau,\, \Omega_S^a(N)\right) = \prod_{j=1}^{n} \mathbf{P}(n_j - n_{j-1} \in \tau). \tag{5.62}$$

By Theorem A.7, $\mathbf{P}(N \in \tau) \overset{N \to \infty}{\sim} \widetilde{L}(N)/N^{1-\alpha}$ for a suitable slowly varying function $\widetilde{L}(\cdot)$. If $\alpha = 1$ then $\widetilde{L}(N) \overset{N \to \infty}{\longrightarrow} 0$. Therefore for every probability $K(\cdot)$ (such that $m_K = \infty$ and $\alpha > 0$) there exists $c > 0$ and $\zeta \in (0,1)$ such that $\mathbf{P}(N \in \tau) \geq c/N^\zeta$. Therefore

$$Z_{N,\widetilde{\omega},\lambda}^{\mathbf{c}} \geq (\exp(\lambda) - 1)^n \sum_{\substack{A:\, A \subset E_N \\ |A| = n}} \prod_{j=1}^{n} \frac{c}{(n_j - n_{j-1})^\zeta}. \tag{5.63}$$

Since, by Jensen inequality, we have

$$\prod_{j=1}^{n} \frac{1}{(n_j - n_{j-1})^\zeta} = \exp\left(-n\zeta \frac{1}{n} \sum_{j=1}^{n} \log(n_j - n_{j-1})\right)$$

$$\geq \exp\left(-\zeta n \log(N/n)\right), \tag{5.64}$$

we arrive at

$$Z^c_{N,\widetilde{\omega},\lambda} \geq (\exp(\lambda) - 1)^n \, |\{A : A \subset E_N, |A| = n\}| \, c^n \left(\frac{n}{N}\right)^{\zeta n}. \qquad (5.65)$$

Let us now choose $n = \lfloor |E_N|/M \rfloor$ for some $M \in \mathbb{N}$ to be chosen below. Remark that $|\{A : A \subset E_N, |A| = n\}| \geq M^n$ and we have

$$Z^c_{N,\widetilde{\omega},\lambda} \geq \left(cM \, (\exp(\lambda) - 1) \, (n/N)^\zeta\right)^n. \qquad (5.66)$$

But since $|E_N| \geq \delta N$, then $n/N \geq \delta/(2M)$ for N sufficiently large and the term between parentheses can be made larger than 1 by choosing M sufficiently large (recall that $\zeta < 1$). This completes the proof. $\qquad \square$

Proof of Theorem 5.9. First of all we see immediately that the estimate in Lemma 5.10 shows that (5.56) implies (5.57). Equation (5.57) implies (5.58): write

$$\log Z^a_{N,\widetilde{\omega},\lambda} = \log \left(\frac{Z^a_{N,\widetilde{\omega},\lambda}}{Z^a_{N,\widetilde{\omega},0}}\right) = \int_0^\lambda \mathbf{E}^a_{N,\widetilde{\omega},\rho} \left[\sum_{n \in \tau : \, n \leq N} \widetilde{\omega}_n\right] d\rho, \qquad (5.67)$$

and since the expectation in the right-hand side is non-decreasing in ρ, by applying (5.57) we get

$$\mathbf{E}^a_{N,\widetilde{\omega},\lambda} \left[\sum_{n \in \tau : \, n \leq N} \widetilde{\omega}_n\right] \geq \frac{1}{\lambda} \log Z^a_{N,\widetilde{\omega},\lambda} \geq \mathrm{const.}\, N. \qquad (5.68)$$

Since of course $\sum_{n \in \tau : \, n \leq N} \widetilde{\omega}_n \leq \mathcal{N}_N(\tau)$, (5.58) follows.

We are left with showing that (5.58) implies (5.56). Let us therefore suppose that $\lim_{N \to \infty} (1/N) \sum_{n=1}^N \widetilde{\omega}_n = 0$. This immediately implies that

$$\lim_{N \to \infty} \frac{1}{N} \log \mathbf{E} \left[\exp\left(\lambda \sum_{n \in \tau : \, n \leq N} \widetilde{\omega}_n + \beta \mathcal{N}_N(\tau)\right) ; \Omega^a_S(N)\right] = \mathrm{F}(\beta), \qquad (5.69)$$

where we have simply used that $\sum_{n \in \tau : \, n \leq N} \widetilde{\omega}_n \leq \sum_{n=1}^N \widetilde{\omega}_n$ and $\mathrm{F}(\beta)$ is the free energy of the homogeneous pinning model of Chapter 2 to which we refer. Since the underlying free process is null recurrent, by Theorem 2.4 we have that $\mathrm{N}(\beta) = \mathrm{F}'(\beta) \xrightarrow{\beta \searrow 0} 0$. This guaranties also the differentiability of $\mathrm{F}(\cdot)$ in 0 and the value of this derivative is zero: by taking the derivative with respect to β of both sides of (5.69) and by exchanging the derivative with the limit (by convexity and regularity) we find

$$\lim_{N \to \infty} \mathbf{E}^a_{N,\tilde{\omega},\lambda} \left[\frac{\mathcal{N}_N(\tau)}{N} \right] = \mathrm{F}'(0) = 0, \tag{5.70}$$

which negates (5.58) and the proof is complete.

Theorem 5.9
□

5.7 Bibliographic Complements

Complements on Section 5.2

As already pointed out Theorem 5.2 is the disordered version of Proposition 3.8. A more general version of the inequality (5.9) of Theorem 5.2 is the main result in [Alexander and Sidoravicius (2006)]. The result is more general because it deals with the most general class of return probabilities, but on the other hand it is not quantitative. Another quantitative proof, but restricted to the simple random walk set-up may be found in [Pétrélis (2005)]. Our technique of proof in a sense borrows the initial step of the argument in [Alexander and Sidoravicius (2006)], that is the targeting of *certain regions*, but it is then rather different and it has been inspired by the lower bound techniques in [Bolthausen and den Hollander (1997)], see Section 6.1. The proof that we present can be generalized to a much larger class of return probabilities (but of course the explicit bounds would not stay the same).

Complements on Section 5.3

A complete proof of Proposition 5.5 can be found in [Caravenna and Giacomin (2005)]: the proof of the fact that $G(\cdot)$ is a gradient if the rightmost term in (5.30) behaves like $\exp(o(N))$ is in a version of the paper that can be found on the webpages of the authors (and on arXiv.org). In [Caravenna and Giacomin (2005)] it is proven only that $\sup_\omega \left| \sum_{n=1}^N G(\theta\omega) \right| = o(N)$, which suffices to conclude. We stress however that the result in [Caravenna and Giacomin (2005)] works beyond the case of linear chains and deals also with higher dimensional models like surface models. The *Morita* approach has been initiated in [Morita (1966)] and developed in particular in [Kühn (1996)]. It has been applied to copolymers for example in [Iliev et al. (2005)] and [Iliev et al. (2004)].

Complements on Section 5.4

It is natural to link the main result of this section with the celebrated paper [Aizenman and Wehr (1990)], which covers in particular the two-dimensional disordered Ising model (see also [Bovier and Külske (1996)] for an application of the argument to $(2+1)$-dimensional interface models). In spite of the fact that the two results are close, we stress that the arguments of proof are completely different. In [Aizenman and Wehr (1990)] the key issue is the competition between bulk disorder and the boundary effect: this competition is analyzed by a subtle argument based on the Central Limit Theorem, *i.e.* typical fluctuations. In our case the boundary plays no role and the argument plays on large deviations, *i.e.* atypical events.

A more general version of Theorem 5.6 is proven in [Giacomin and Toninelli (2006a)]: besides more general $K(\cdot)$ (in this case the polynomial decay is important, but one needs only a lower bound) it covers also the case of ω_1 with a general bounded distribution. The argument that we give here is a *direct* rigorous version of the heuristic argument given in [Giacomin and Toninelli (2006c)]. Notice that the argument we give here does not go through for bounded disorder, since the relative entropy is unbounded: in that case one has rather to use the standard tilting procedure, *i.e.* modifying the mean of the disorder by introducing a chemical potential, but the difficulty is then that the disorder is *globally* modified (it is not just a shift, like in the cases we treat here). One then has to play on controlling the difference between free energies with (slightly) different disorders and this is done by introducing a constraint on the contact density. As a matter of fact, the argument given in [Giacomin and Toninelli (2006a)] is a bit more involved precisely because in the selected rare regions only the paths with a suitable contact density are kept. It is reasonable that Theorem 5.6 hold for even more general distribution of ω_1, but the generalization of the technique of proof does not appear to be immediate. The technique should also generalize to (weakly) correlated environments, but this direction has not been investigated.

The smoothing effect of disorder in pinning models is a controversial issue in the physical literature, at least for $\alpha > 1$. These are precisely the return probabilities considered to be of relevance for the DNA denaturation modeling (see Section 1.4 and relative bibliographic complements). Relevant papers that attack the problem of regularity in disordered Poland–Scheraga models are in particular [Cule and Hwa (1997)], [Tang and Chaté (2001)], [Blossey and Carlon (2003)], [Schäfer (2005)], [Coluzzi (2005)],

[Garel and Monthus (2005a)] and [Garel and Monthus (2005b)]. In the latter two it is claimed, based on numerical and theoretical arguments, that the transition is of first order for $\alpha > 1$: this is in contrast with the main result of this section. There seems to be instead general agreement on the fact that the transition is of at least second order if $\alpha < 1$ (see the complements to the next section for more on this). We take this occasion to point out that real DNA sequences display long range correlation dependence (long range correlations in DNA base sequences has been studied in depth, starting with the seminal paper [Peng *et al.* (1992)], and it still represents an active field of research).

Complements on Section 5.5

The version of Harris' argument [Harris (1974)] given here has been communicated to me by F. L. Toninelli. The claim that at $\alpha = 1/2$ the noise is marginal is (implicitly and explicitly) contained in several papers. As a matter of fact, there is essentially full agreement in the physical literature that the disorder is irrelevant (respectively relevant) if $\alpha < 1/2$ (respectively $\alpha > 1/2$), [Forgacs *et al.* (1986)] and [Derrida *et al.* (1992)]. Note that this in particular would imply that the critical behavior of the quenched model is the same of the critical behavior on the non-disordered model for $\alpha < 1/2$, so that the transition is of order two or larger, see Figure 5.4. The controversial issue is about the case $\alpha = 1/2$ and $L(\cdot)$ trivial, that is the case of random walks, where the disorder is *marginal* and the two papers we just mentioned do not agree on whether the noise is *marginally relevant* or *marginally irrelevant*. In particular in [Forgacs *et al.* (1986)] it is claimed that for β sufficiently small then quenched and annealed critical points coincide, while in [Derrida *et al.* (1992)] an exact computation is carried out on a simplified model and the claim is that quenched and annealed points differ by a leading order correction of the type $c_1 \exp(-c_2/\beta^2)$, *i.e. beyond all orders*. For updates and recent literature on this unresolved issue we refer to [Tang and Chaté (2001)].

Complements on Section 5.6

The section is adapted from [Janvresse *et al.* (2005)], where the analysis in dimension $(1+1)$ is restricted to symmetric random walk with finite variance and therefore to $\alpha = 1/2$. But also the case of walks in dimension $(1+2)$ and of $(2+1)$ dimensional effective interface models are considered (obtaining the analogous result), the most delicate case being the one of walks in dimension $(1+2)$. Here we have not treated the case of $\alpha = 0$: this is because, with reference to the end of the proof of Lemma 5.10, we would have to choose $\zeta = 1$ and the argument we give is not sufficient to conclude. However the result holds also in that case since the case $\zeta = 1$ is treated in [Janvresse *et al.* (2005)] when dealing with the $(1+2)$ dimensional case. Note that, as it is natural, the argument in [Janvresse *et al.* (2005)] are based on the classical Local Limit Theorem, see (A.4), precisely in order to estimate the Green's function of the walk, while in our case this is replaced by mass renewal function estimates.

Chapter 6

Disordered Copolymers and Selective Interfaces: The Phase Diagram

In this section we obtain a number of estimates on the free energy of the disordered copolymer model (Section 1.7). Except for Section 6.3.2, no adsorption term will be included, so throughout the chapter $\hat{\omega} = \omega$ (with the noted exception).

Copolymers are naturally introduced as based on $(1 + 1)$-dimensional random walk models. In Section 6.3.1 we will argue that this is a rather reasonable model also for higher dimensional situation, in particular the case of a copolymer in three dimensions close to a (two-dimensional) flat interface separating two selective solvents. So $\alpha = 1/2$ is definitely the most relevant case, but we will (almost) systematically treat the case $\alpha \in (0, 1)$ and, when of interest, also the cases $\alpha \geq 1$. On the other hand we will restrict, for sake of conciseness, to recurrent renewals. These return exponents appear in a random walk context for example when one adds suitable (height dependent) drifts toward the interface. But of course, one can introduce the copolymer model simply by giving the renewal process τ of the returns and the signs of the excursions (independent fair coin tossing). In this section we assign at random also the signs of the excursions of length one. This avoids singling out these excursions and somewhat lightens the notations. The main arguments of the first two sections of this chapter go through with straightforward changes for other choices of the sign: this is due to the fact that these arguments are based on *long excursions*. However the dependence on the choice of the sign for disordered systems is still not understood and it is a particular case of the problem of understanding the superposition of copolymer and pinning interactions (the so called copolymer with adsorption model). In Section 6.3 the role of the choice of the sign will be discussed in (some) depth.

Let us recall that $\omega(j, k] = \sum_{n=j+1}^{k} \omega_n$ (and $\omega(j, j] = 0$) and let us spell out the partition function of the model, *cf.* (1.60): with $s_n :=$ sign$((S_{n-1}, S_n)) \in \{-1, +1\}$, choosing s_n by tossing a fair coin if both S_{n-1} and S_n are zero, we have

$$
\widetilde{Z}^{\mathfrak{f}}_{N,\omega} = \mathbf{E}\left[\exp\left(\lambda\sum_{n=1}^{N}(\omega_n + h)s_n\right)\right]
$$
$$
= \mathbf{E}\left[\widetilde{R}_{N,\omega}(\tau)\prod_{j=1}^{\mathcal{N}_N(\tau)}\exp\left(\psi\big(\lambda\omega(\tau_{j-1}, \tau_j] + \lambda h(\tau_j - \tau_{j-1})\big)\right)\right],
$$
$$(6.1)$$

with $\psi(t) = \log\cosh(t)$ and

$$
\widetilde{R}_{N,\omega}(\tau) = \cosh\left(\lambda\omega(\tau_{\mathcal{N}_N(\tau)}, N] + \lambda h(N - \mathcal{N}_N(\tau))\right), \qquad (6.2)
$$

is the contribution of the last (incomplete) excursion. Alternatively, *cf.* (1.62):

$$
Z^{\mathfrak{f}}_{N,\omega} = \mathbf{E}\left[\exp\left(-2\lambda\sum_{n=1}^{N}(\omega_n + h)\Delta_n\right)\right]
$$
$$
= \mathbf{E}\left[R_{N,\omega}(\tau)\prod_{j=1}^{\mathcal{N}_N(\tau)}\exp\left(\varphi\big(\lambda\omega(\tau_{j-1}, \tau_j] + \lambda h(\tau_j - \tau_{j-1})\big)\right)\right],
$$
$$(6.3)$$

with $\varphi(t) = \log((1 + \exp(-t))/2)$ and

$$
R_{N,\omega}(\tau) = \exp\left(\varphi\big(\lambda\omega(\tau_{\mathcal{N}_N(\tau)}, N] + \lambda h(N - \mathcal{N}_N(\tau))\big)\right). \qquad (6.4)
$$

As pointed out in Section 1.6.2, $\widetilde{Z}^{\mathfrak{f}}_{N,\omega} = \exp\left(\lambda\sum_{n=1}^{N}(\omega_n + h)\right)Z^{\mathfrak{f}}_{N,\omega}$. So by the Strong Law of Large Numbers (Theorem A.1) and by the existence of the free energy (Theorem 4.1)

$$
\mathrm{F}(\lambda, h) = \lim_{N\to\infty}\frac{1}{N}\log Z^{\mathfrak{f}}_{N,\omega} = -\lambda h + \lim_{N\to\infty}\frac{1}{N}\log\widetilde{Z}^{\mathfrak{f}}_{N,\omega}, \qquad (6.5)
$$

and the limits are to be intended in the a.s. or in the L^1 sense. Of course the same statement and analogous formulas hold for the constrained endpoint case.

Recall also that for the copolymer λ and h are chosen (without loss of generality) non-negative. Set for $m > 0$, $\lambda > 0$ and $-2\lambda m \notin \partial D_{\mathrm{M}}$ (*cf.*

Definition 1.16)

$$h^{(m)}(\lambda) := \frac{1}{2m\lambda} \log M(-2m\lambda). \tag{6.6}$$

The definition is then extended by continuity to 0, that is $h^{(m)}(0) := 0$, and by continuity from the left when $-2m\lambda$ is at the boundary of D_M, so that $h^{(m)}(\cdot)$ is lower semicontinuous (see Figure 6.1). Note that $\lim_{\lambda \searrow 0} dh^{(m)}(\lambda)/d\lambda = m$.

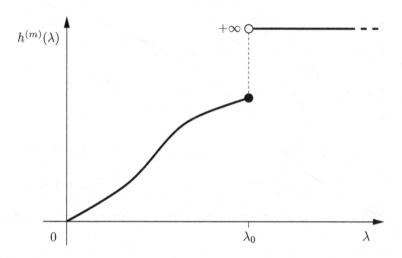

Fig. 6.1   When ω_1 is only locally exponentially integrable and the left-tail of its distributions does not decay fast enough, so that the interval D_M is bounded from the left and $-2m\lambda_0 \in \partial D_M$ for some $\lambda_0 > 0$, it may happen that $\lim_{\lambda \nearrow \lambda_0}(1/2m\lambda) \log M(-2m\lambda)$ is finite, while $(1/2m\lambda_0) \log M(-2m\lambda_0) = \infty$. In this case, and only in this case, $h^{(m)}(\lambda_0) \neq (1/2m\lambda_0) \log M(-2m\lambda_0)$.

One of the main results proven in this chapter is the following:

Theorem 6.1

(1) There exists an increasing function $h_c : [0, \infty) \to [0, \infty]$, lower semi-continuous on $[0, \infty)$ and continuous on $[0, \sup\{t : h_c(t) < \infty\})$, such that $h_c(0) = 0$ and such that $\lambda \mapsto \lambda h_c(\lambda)$ is a convex function, that characterizes $\mathcal{L}$ (and therefore $\mathcal{D}$):

$$\mathcal{L} = \{(\lambda, h) : h < h_c(\lambda)\}. \tag{6.7}$$

(2) We have the following bounds:

$$h^{(1/(1+\alpha))}(\lambda) \leq h_c(\lambda) \leq h^{(1)}(\lambda), \qquad (6.8)$$

for every λ.

By increasing we mean (strictly) increasing on the set in which the function is bounded.

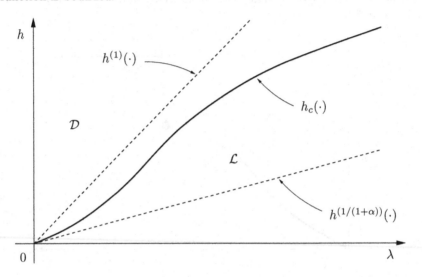

Fig. 6.2   The critical curve and the bounds of (6.8). We are considering the case of $\omega_1 \sim \mathcal{N}(0,1)$, where both lower and upper bounds are straight lines. Is the critical curve a straight line too? This is an open problem, but we know that the critical curve is increasing and that $\lambda \mapsto \lambda h_c(\lambda)$ is convex. We take this occasion to also point out that for $\alpha = 0$ we have $h_c(\cdot) \equiv h^{(1)}(\cdot)$.

In this chapter we will explore more than one technique obtaining several estimates that go (substantially) beyond Theorem 6.1. In particular, Theorem 6.1 is for example an immediate consequence of Lemma 6.2 and Theorem 6.5.

6.1   Copolymers and Homogeneous Localization

This section is devoted to applying the homogeneous localization strategy, already exploited in Section 5.2 in the context of disordered pinning models. But let us first prove the *soft part* of Theorem 6.1, namely:

Lemma 6.2 *Let us assume that there exist $m > 0$ and $\varepsilon > 0$ such that $\{(\lambda, h) : \lambda < \varepsilon, \ h = m\lambda\} \subset \mathcal{D}$ and let us assume also that $\mathrm{F}(\lambda, 0) > 0$ for $\lambda \in (0, \varepsilon)$. Then part (1) of Theorem 6.1 holds.*

The apparently strange conditions are simply requiring (1) that the polymer is localized if $h > 0$, as long as $\lambda > 0$ (in fact knowing localization for $\lambda \in (0, \varepsilon)$, by convexity of $\mathrm{F}(\cdot, h)$ and the fact that $\mathrm{F}(0, 0) = 0$, implies localization on all the semi-axis), and (2) that the system is delocalized on a line with sufficiently large slope, at least close to the origin. By monotonicity in h this implies delocalization above this line.

Proof. Let us consider the model with energy $-2 \sum_n (\lambda \omega_n + u) \Delta_n$, $u \geq 0$, and let us call $\mathrm{G}(\lambda, u)$ the corresponding free energy. Note that this is just a change of variables, at least if $\lambda > 0$, and we have $\mathrm{F}(\lambda, u/\lambda) = \mathrm{G}(\lambda, u)$, so in particular $\mathrm{G}(\cdot, \cdot) \geq 0$. By hypothesis $\mathrm{G}(\lambda, 0) > 0$ for every $\lambda > 0$ and $\mathrm{G}(\lambda, u) = 0$ for $u \geq m\lambda^2$, at least if $\lambda < \varepsilon$. Note also that $\mathrm{G}(\cdot, \cdot)$ is a convex function, that $\mathrm{G}(\lambda, \cdot)$ is non-increasing and that $\mathrm{G}(0, \cdot) = 0$. Consider the level set $\mathbf{L}_0 := \{(\lambda, u) : \mathrm{G}(\lambda, u) = 0\}$, which, since it coincides with $\{(\lambda, u) : \mathrm{G}(\lambda, u) \leq 0\}$, is a convex set, see Appendix A.1.1. Therefore if we set $u_c(\lambda) := \inf\{u : \mathrm{G}(\lambda, u) = 0\}$, then $u_c(\cdot)$ is bounded in a neighborhood of the origin (in fact we have a quadratic upper bound), with $u_c(0) = 0$, and $\{(\lambda, u_c(\lambda)) : \lambda > 0, \ u_c(\lambda) < \infty\}$ is a subset of the boundary of $\mathbf{L}_0$. Therefore $\lambda \mapsto u_c(\lambda)$ is convex and therefore continuous except at most in one point, the point at which it diverges (if it does diverge): at this point the function is lower semicontinuous.

Now we set $h_c(\lambda) := u_c(\lambda)/\lambda$. Note that, by change of variables, for $\lambda > 0$ we have $h_c(\lambda) = \inf\{h : \mathrm{F}(\lambda, h) = 0\}$ and we readily see that $h_c(\cdot)$ is continuous at all points at which it is bounded: the continuity in zero follows since $u_c(\cdot)$ has a quadratic upper bound near the origin.

The fact that $\lambda \mapsto h_c(\lambda)$ is non-decreasing is an immediate consequence of the fact that $\mathrm{F}(\cdot, h)$ is non-decreasing, which in turn follows from the fact that $\mathrm{F}(0, h) = 0$ and from the convexity and non-negativity of $\mathrm{F}(\cdot, h)$. But we want to prove *strict* monotonicity and the argument goes as follows. Choose $\lambda' > \lambda$ and assume $h_c(\lambda') < \infty$ (otherwise there is nothing to prove). By convexity we have $u_c(\lambda') \geq \partial_+ u_c(\lambda)(\lambda' - \lambda) + u_c(\lambda)$, with ∂_+ the derivative from the right, *cf.* Appendix A.1.1. Therefore

$$h_c(\lambda') - h_c(\lambda) \geq (\lambda \partial_+ u_c(\lambda) - u_c(\lambda)) \left(\frac{1}{\lambda} - \frac{1}{\lambda'} \right). \tag{6.9}$$

Monotonicity follows if we show that $\lambda\partial_+ u_c(\lambda) - u_c(\lambda) > 0$. But, by convexity and the fact that $u_c(0) = 0$, if $\lambda\partial_+ u_c(\lambda) - u_c(\lambda) = 0$ for some $\lambda > 0$, then $\partial_+ u_c(\cdot) \equiv c \geq 0$ on $(0, \lambda]$. The case $c = 0$ is in contrast with the fact that $u_c(\lambda) > 0$ for $\lambda > 0$ and the case $c > 0$ is in contrast with the quadratic upper bound on $u_c(\cdot)$ close to the origin. $\square$

The main result of this section is:

Theorem 6.3   *For every $\alpha > 0$ there exist two positive constant $c_1 < c_2$ such that*

$$c_1\lambda^2 \leq \mathrm{F}(\lambda, 0) \leq c_2\lambda^2, \qquad (6.10)$$

for every $\lambda \leq 1$. Moreover if $m_K < \infty$

$$\lim_{\lambda \searrow 0} \frac{1}{\lambda^2}\mathrm{F}(\lambda, 0) = 1/2. \qquad (6.11)$$

Both c_1 and c_2 can be made rather explicit. In fact c_2 is very explicit (and independent of α!), see (6.12). On the other hand c_1 depends on α and an explicit bound on c_1 can be extracted from (6.17).

Theorem 6.3 can be easily upgraded to a statement for $h > 0$ (the details are in Section 6.4). Note that Theorem 6.3 proves one of the two hypotheses under which Lemma 6.2 holds (and the upgrade in Section 6.4 proves the second hypothesis, which is however also a consequence of the bounds that we prove in the next section).

Proof. The upper bound is just the annealed bound:

$$\mathrm{F}(\lambda, 0) \leq \lim_{N\to\infty} \frac{1}{N}\log \mathbb{E}\mathbb{E}\left[\exp\left(\lambda\sum_{n=1}^{N}\omega_n s_n\right)\right]$$

$$= \lim_{N\to\infty} \frac{1}{N}\log \mathbb{E}\left[\exp\left(\sum_{n=1}^{N}\log \mathrm{M}\left(\lambda s_n\right)\right)\right] \qquad (6.12)$$

$$= \max_{s=\pm 1} \mathrm{M}(s\lambda) \stackrel{\lambda\searrow 0}{\sim} \frac{1}{2}\lambda^2.$$

The equality in the last line has been obtained by observing that on one hand $\max_{s=\pm 1} \mathrm{M}(s\lambda)$ is clearly an upper bound for the term in the second line of (6.12), but also a lower bound as we see by restricting the expectation to the event $\tau_1 > N$ and by choosing the proper sign of the excursion.

For the lower bound we apply the homogeneous localization strategy. The notation is the one of Appendix B.1. We start by claiming

$$F(\lambda, 0) \geq \sup_{b > 0} \left(\frac{1}{m_{K_b}} \sum_n K_b(n) \mathbb{E}\left[\psi(\lambda \omega(0, n))\right] - s(b) \right). \qquad (6.13)$$

A proof can be found below. We now go on with the main body of the argument. Choose $\alpha \in (0, 1)$ and set $b = \kappa \lambda^2$, $\kappa > 0$, and consider the small λ behavior of the term between parentheses in the right-hand side of (6.13). By Proposition B.2 and by (B.14) we get

$$\frac{1}{m_{K_b}} \sum_n K_b(n) \mathbb{E}\left[\psi(\lambda \omega(0, n))\right] - s(b) \overset{\lambda \searrow 0}{\sim}$$

$$\frac{(1 - \alpha) b^{1 - \alpha}}{\Gamma(2 - \alpha) L(1/b)} \sum_n \frac{L(n)}{n^{1 + \alpha}} \exp(-bn) \mathbb{E}\left[\psi(\lambda \omega(0, n))\right] - \frac{1 - \alpha}{\alpha} b. \qquad (6.14)$$

Let us remark that for $t > 0$ by the Central Limit Theorem (see (A.3)), applied in conjunction with the uniform integrability bound $\sup_\lambda \mathbb{E}[|\lambda \omega(0, \lfloor t/b \rfloor)|^2] < \infty$, we have

$$\lim_{\lambda \searrow 0} \mathbb{E}\left[\psi(\lambda \omega(0, \lfloor t/b \rfloor))\right] = \mathbb{E}\left[\psi\left(Z\sqrt{t/\kappa}\right)\right], \qquad (6.15)$$

where $Z \sim \mathcal{N}(0, 1)$. Observe that for every $\varepsilon \in (0, 1)$

$$\frac{b^{1 - \alpha}}{L(1/b)} \sum_{\substack{n: \\ \varepsilon < bn < 1/\varepsilon}} \frac{L(n)}{n^{1 + \alpha}} \exp(-bn) \mathbb{E}\left[\psi(\lambda \omega(0, n))\right] \overset{\lambda \searrow 0}{\sim}$$

$$b \int_\varepsilon^{1/\varepsilon} \frac{\exp(-t)}{t^{1 + \alpha}} \mathbb{E}\left[\psi\left(Z\sqrt{t/\kappa}\right)\right] \, dt, \qquad (6.16)$$

where we have also used the uniform converge property (A.18) of slowly varying functions. Note that, since there exists $c > 0$ such that $\log \cosh(t) \leq ct^2$ for every t, $\mathbb{E}\left[\psi\left(Z\sqrt{t/\kappa}\right)\right] \leq ct/\kappa$. By letting ε tend to zero we get to the bound:

$$\liminf_{\lambda \searrow 0} \frac{1}{\lambda^2} F(\lambda, 0) \geq \kappa \frac{1 - \alpha}{\Gamma(2 - \alpha)} \int_0^\infty \frac{\exp(-t)}{t^{1 + \alpha}} \mathbb{E}\left[\psi\left(Z\sqrt{t/\kappa}\right)\right] \, dt - \kappa \frac{1 - \alpha}{\alpha}. \qquad (6.17)$$

We are left with showing that we can choose $\kappa > 0$ such that the right-hand side is positive. By restricting the integral (for example) to $t \in [1, 2]$ and by observing that $\mathbb{E}\left[\psi(Zr)\right] \geq r/2$ for r sufficiently large, as it follows immediately from $\psi(t) \geq |t| - \log 2$, then one sees that the first term in the right-hand side is bounded below by (a positive constant times) $\sqrt{\kappa}$, at least for κ sufficiently small. And it is precisely by choosing κ sufficiently small that the right-hand side of (6.17) becomes positive.

The claim (6.13) follows from (A.10) applied to $\widetilde{Z}_{N,\omega}^{\mathrm{f}}$ (in the form of the second line in (6.1)), by choosing as comparison process the renewal with inter-arrival law $K_b(\cdot)$ ($\mathbf{P}_b$ is its law). Proposition B.1 takes care of the entropy term. We are therefore left with showing two facts: the first is that

$$\lim_{N\to\infty} \frac{1}{N}\mathbf{E}_b \sum_{j=1}^{\mathcal{N}_N(\tau)} \mathbb{E}\left[\psi\left(\lambda\omega(\tau_{j-1},\tau_j])\right)\right] = \frac{1}{m_{K_b}} \sum_n K_b(n)\mathbb{E}\left[\psi(\lambda\omega(0,n])\right],$$
(6.18)

which is in turn a consequence of

$$\lim_{N\to\infty} \frac{1}{N}\mathbf{E}_b \sum_{j=1}^{N} \mathbb{E}\left[\psi\left(\lambda\omega(\tau_{j-1},\tau_j])\right)\right] = \sum_n K_b(n)\mathbb{E}\left[\psi(\lambda\omega(0,n])\right], \quad (6.19)$$

since $\lim_{N\to\infty} \mathcal{N}_N(\tau)/N = m_{K_b}$, $\mathbf{P}_b(\mathrm{d}\tau)$–a.s. (which follows from the Strong Law of Large Numbers, *cf.* (A.31)). The second assertion to be proven is that $\mathbb{E}\mathbf{E}_b\left[\widetilde{R}_{N,\omega}(\tau)\right]/N$ vanishes as $N \to \infty$. This point however is immediate, because $|\widetilde{R}_{N,\omega}| \le c_1 + \lambda \sum_{n=\tau_{\mathcal{N}_N(\tau)}}^{N} |\omega_n|$. By taking the $\mathbb{P}$–expectation one is left with estimating the expectation of the backward recurrence time, which is of order 1, so $\mathbb{E}\mathbf{E}_b\left[\widetilde{R}_{N,\omega}(\tau)\right]/N = O(1/N)$.

The proof of (6.19) is quick too. Note that for $k < m$

$$\mathbb{E}\left[\psi\left(\lambda\omega(k,m])\right)\right] = \mathbb{E}\left[\psi\left(\lambda\omega(0,m-k])\right)\right], \qquad (6.20)$$

and (6.19) follows immediately from the law of large numbers applied to the IID sequence $\{\mathbb{E}\left[\psi\left(\lambda\omega(0,\tau_j-\tau_{j-1}])\right)\right]\}_j$.

The proof in the case of $m_K < \infty$ is substantially easier, since it does not require the localization procedure ($b = 0$ in (6.14)). However we want to get the precise estimate (6.11): the upper bound is already in (6.12). For the lower bound we use Jensen inequality (namely the localization procedure with $b = 0$) to get

$$\frac{1}{N}\mathbb{E}\log\widetilde{Z}_{N,\omega}^{\mathrm{f}} \ge \frac{1}{N}\mathbf{E}\left[\sum_{j=1}^{\mathcal{N}_N(\tau)} \mathbb{E}\psi\left(\lambda\omega(\tau_{j-1},\tau_j])\right)\right]$$
$$\stackrel{N\to\infty}{\to} \frac{1}{m_K}\mathbb{E}\mathbb{E}\psi\left(\lambda\omega(0,\tau_1])\right),$$
(6.21)

where we have used (A.31). Observe now that by the integrability properties of ω we have that $\mathbb{E}\psi\left(\lambda\omega(0,n]\right)/\lambda^2$ converges as $\lambda \searrow 0$ to $n/2$ for every n. Therefore by Fatou's Lemma from (6.21) we obtain

$$\liminf_{\lambda \searrow 0} F(\lambda, 0) \geq \frac{1}{m_K} \sum_n \frac{n}{2} K(n) = \frac{1}{2}, \qquad (6.22)$$

and the proof of (6.11) is complete.

We leave working out the details for the case of $\alpha = 1$ with $m_K = \infty$ to the reader. $\qquad\square$

Remark 6.4 *It is interesting to note that from (6.17) one directly extracts that*

$$\lim_{\alpha \nearrow 1} \liminf_{\lambda \searrow 0} \frac{1}{\lambda^2} F(\lambda, 0) = 1/2. \qquad (6.23)$$

This is obtained by choosing $\kappa := \kappa(\alpha)$ such that $\kappa(\alpha) \to \infty$ and $(1 - \alpha)\kappa(\alpha) \to 0$, as $\alpha \nearrow 1$, and by recalling that $\Gamma(1 - \alpha) \sim 1/(1 - \alpha)$ in the same limit. Note in fact that both terms contain the (vanishing) term $\kappa(1 - \alpha)$ and, since $\Gamma(2 - \alpha) \to 1$, the main role is played by the integral term. Use then $\mathbb{E}[\psi(Za)] \overset{a \searrow 0}{\sim} a^2/2$ to conclude.

6.2 Rare Stretch Strategy for Localization Estimates

The heart of this section is an approach to localization estimates that is, in a sense, *orthogonal* to the homogeneous localization strategy. The result is the following:

Theorem 6.5 *We have the inclusions*

$$\left\{(\lambda, h) : \ h \geq h^{(1)}(\lambda)\right\} \subset \mathcal{D}, \qquad (6.24)$$

and

$$\left\{(\lambda, h) : \ h < h^{(1/(1+\alpha))}(\lambda)\right\} \subset \mathcal{L}. \qquad (6.25)$$

In different terms, this statement is just (6.8) without passing through defining $h_c(\cdot)$.

Proof. The proof of (6.24) follows from the annealed bound applied to the first line in (6.3):

$$\mathrm{F}(\lambda, h) \le \lim_{N \to \infty} \frac{1}{N} \log \mathbb{E} Z_{N,\omega}^{\mathrm{f}} = \lim_{N \to \infty} \frac{1}{N} \log \mathbb{E}\left[\exp\left(\widetilde{\beta} \sum_{n=1}^{N} \Delta_n\right)\right],$$

$$\tag{6.26}$$

where $\widetilde{\beta} := \log \mathrm{M}(-2\lambda) - 2\lambda h$. The last term is equal to zero if (and only if) $\widetilde{\beta} \le 0$ and this is precisely the condition $h \ge h^{(1)}(\lambda)$, at least if -2λ is not on the boundary of $D_{\partial \mathrm{M}}$, since in that point $h^{(1)}(\lambda)$ may differ from $\log \mathrm{M}(-2\lambda)/(2\lambda)$. However, since $\mathcal{D}$ is a closed subset of $[0, \infty)^2$, (6.24) is implied anyway.

We turn now to the proof of (6.25). Let us first explain the idea of the proof. If the transition is of order higher than one (which we are able to prove under suitable conditions, *cf.* Section 5.4 and Section 6.4 (but we are not going to use it in this proof), then if h is smaller but very close to $h_c(\lambda)$ the polymer is mostly in the upper half plane. Of course there is still a positive density of monomers in the lower half plane, since the polymer is localized, but this density is close to zero. The trajectories should therefore resemble the trajectory drawn in Figure 6.3: sparse visits in the lower half plane. It is reasonable to believe that these visits are paid if the monomer charges are atypically negative, since there is no energetic gain in visiting the lower half plane, unless $\omega_n < -h$.

Let us make this idea precise by employing a discretization of step ℓ, $1 \ll \ell \ll N$ and by calling $Q_j(\omega)$ the total charge in the jth interval (that we call stretch), that is $Q_j(\omega) := \omega((j-1)\ell, j\ell] + h\ell$. We may safely assume that $N/\ell \in \mathbb{N}$. Of course $\{Q_j\}_j$ is an IID family of random variables. Let us set for $q > 0$

$$p(\ell) := \mathbb{P}(Q_1 \le -q\ell) \overset{\ell \to \infty}{\asymp} \exp\left(-\ell\Sigma(-q-h)\right). \tag{6.27}$$

The asymptotic behavior follows from (A.8). The sequence of Bernoulli random variables $\{Y_j(\omega)\}_j$, $Y_j(\omega) := \mathbf{1}_{Q_j(\omega) \le -q\ell}$, of parameter $p(\ell)$ describes the location of atypical stretches along the chains. Their distance is typically $\ell/p(\ell) \asymp 1/p(\ell)$, and therefore exponentially far with respect to the length of an atypical stretch.

The strategy now consists in estimating the partition function from below by considering only the trajectories that visit the lower half plane exactly at all the atypical stretches. Notice that computing the partition function restricted to these trajectories is particularly easy, because this restricted partition function factorizes (and therefore the contributions to the free energy add up).

With reference to Figure 6.3, the structure is approximately repetitive, so we have a winning strategy if the strategy leads to a positive contribution to the free energy from the very first visit to a stretch. Let us therefore look at the contribution up to the end of the first visit to the lower half plane (let us call $(G+1)\ell$ this quantity): in this section the contribution to the free energy is at least:

$$\log K \left(G\ell\right) + \log \left(K(\ell) \exp(2\lambda q\ell)\right) \approx -(1+\alpha)\Sigma(-q-h)\ell + 2\lambda q\ell, \quad (6.28)$$

where we have neglected the terms $\log \ell$ since they are much smaller than ℓ (and of course also the terms $\log L(\exp(c\ell))$, by the elementary bounds on the slowly varying function $L(\cdot)$). Therefore this strategy leads to a positive contribution to the free energy if $-(1+\alpha)\Sigma(-q-h)+2\lambda q > 0$ and we have still the freedom of optimizing over $q > 0$. In fact we can directly optimize over $q \in \mathbb{R}$ since we are looking for a positive contribution to the free energy and therefore the maximum, if it is positive, will be achieved for $q > 0$. But since the Legendre transform is an involution the localization condition is

$$
\begin{aligned}
0 &< \sup_{q} \left(2\lambda q - (1+\alpha)\Sigma(-q-h)\right) \\
&= -2\lambda h + (1+\alpha) \sup_{q} \left((-2\lambda/(1+\alpha))q - \Sigma(q)\right) \quad (6.29) \\
&= -2\lambda h + (1+\alpha) \log \mathrm{M}\left(-2\lambda/(1+\alpha)\right),
\end{aligned}
$$

and this is the localization condition that we have claimed (note that in the second line we have made the change of variable $q + h \to -q$). To be precise, *cf.* (A.7), (6.29) holds if $-2\lambda \notin \partial D_{\mathrm{M}}$: otherwise in general one has to redefine $\mathrm{M}(\cdot)$ at that point to make $\mathrm{M}(\cdot)$ continuous from the right, and hence lower semicontinuous.

Let us clean up this argument. First of all let us note that if λ is in the interior of the interval D_{M} then there exists $q_0 > 0$ that solves the variational problem (6.29). If instead $\mathrm{M}\left(-2\lambda/(1+\alpha)\right) = \infty$ by choosing q_0 sufficiently large we are assured of a positive contribution. So in any case

$$2\lambda q_0 - (1+\alpha)\Sigma(-q_0 - h) > 0. \quad (6.30)$$

And since $\mathcal{L}$ is an open subset of $[0,\infty)^2$, we need not worry about the value at the boundary of the interval in which $D_{\mathrm{M}} < \infty$.

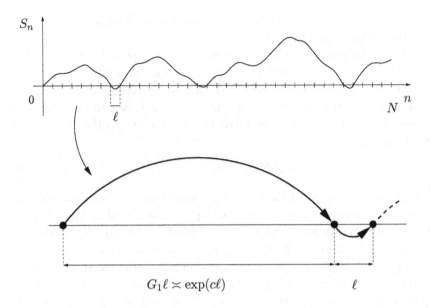

Fig. 6.3 The lower bound strategy is obtained by restricting to polymer trajectories that make negative excursions only in correspondence of rare stretches (three in the figure) of ℓ monomers.

Let us set $A_\ell(\omega) := \{0, N\} \cup \{\ell j : Y_j = 1 \text{ or } Y_{j+1} = 1, \ j < N/\ell\}$ and $\Omega_S(\omega, \ell, N) = \Omega_S(\omega)$ is the ensemble of polymer trajectories such that

(1) $\tau = A_\ell(\omega)$;
(2) $s_n = -1$ if $n \in \cup_{j:\, Y_j = 1}\{(j-1)\ell + 1, j\ell\}$ and $s_n = +1$ otherwise.

See Figure 6.3. Of course $Z^c_{N,\omega} \geq Z^c_{N,\omega}(\Omega_S(\omega))$ and therefore

$$\frac{1}{N} \log Z^c_{N,\omega} \geq 2\lambda \frac{1}{N} |\{j = 1, \ldots, N/\ell : Y_j(\omega) = 1\}| \ell q_0 + \frac{1}{N} \log \mathbf{P}\left(\Omega_S(\omega)\right),$$
(6.31)

where we have simply used the renewal property and the fact that the Hamiltonian is a constant when restricted to the trajectories in $\Omega_S(\omega)$. If we set $\mathcal{N}_Y(\omega) := |\{j = 1, \ldots, N/\ell : Y_j(\omega) = 1\}|$ by the Strong Law of Large Numbers

$$\lim_{N \to \infty} \frac{1}{N} \mathcal{N}_Y(\omega) = \frac{p(\ell)}{\ell},$$
(6.32)

$\mathbb{P}(\,d\omega)$–a.s.. But

$$\log \mathbf{P}\left(\Omega_S(\omega)\right) = \mathcal{N}_Y(\omega)\log(K(\ell)/2) + \sum_{n=1}^{\mathcal{N}_Y(\omega)} \log(K(G_n(\omega)\ell)/2) + O(\log N),$$
(6.33)

where $O(\log N)$ is a bound on the logarithm of the probability of the last excursion, that is of length at most N and $G_n(\omega)$ is the random variable taking value in $\mathbb{N} \cup \{0\}$ that counts the number of ℓ intervals between the successive atypical ℓ intervals (for a precise definition see Section 5.4 where the analogous quantity has been defined in detail). Using the expression of $K(\cdot)$ (for the occasion we set $K(0) := 1$) and the properties of slowly varying functions we easily obtain that if $\varepsilon > 0$ for ℓ sufficiently large we have

$$\frac{1}{N} \sum_{n=1}^{\mathcal{N}_Y(\omega)} \log K(G_n\ell) \geq -\frac{\mathcal{N}_Y(\omega)}{N} \frac{(1+\alpha+\varepsilon)}{\mathcal{N}_Y(\omega)} \sum_{n=1}^{\mathcal{N}_Y(\omega)} \log(\hat{G}_n\ell), \quad (6.34)$$

where $\hat{G}_n := \max(G_n, 1/\ell)$ has been introduced to match with the definition $K(0) = 1$. Jensen inequality implies that

$$\frac{1}{\mathcal{N}_Y(\omega)} \sum_{n=1}^{\mathcal{N}_Y(\omega)} \log(\hat{G}_n\ell) \leq \log\left(\frac{1}{\mathcal{N}_Y(\omega)} \sum_{n=1}^{\mathcal{N}_Y(\omega)} \hat{G}_n\ell\right) \leq \log\left(\frac{N}{\mathcal{N}_Y(\omega)}\right).$$
(6.35)

Combining (6.32)–(6.35) we get

$$\liminf_{N\to\infty} \frac{1}{N} \log \mathbf{P}\left(\Omega_S(\omega)\right) \geq \frac{p(\ell)}{\ell}(1+\alpha+\varepsilon)(\log p(\ell) - \log \ell), \quad (6.36)$$

and for ℓ large $\log \ell$ becomes negligible compared to $\log p(\ell)$. Plugging (6.36) into (6.31) we have

$$\textsc{f}(\lambda, h) \geq p(\ell)\big[2\lambda q_0 - (1+\alpha+\varepsilon)\Sigma(-q_0 - h) + o_\ell(1)\big]. \quad (6.37)$$

Since the term between brackets in the right-hand side converges if we send $\ell \to \infty$ and then $\varepsilon \searrow 0$ to the expression in (6.30), which is positive, we can find ℓ and ε that make the right-hand side of (6.37) positive and we are done. □

6.3   The Weak Coupling Limit: Brownian Motion Model and Universality

This section is expository and it deals with one of the most interesting aspects of the copolymer model. This aspect is however still underdeveloped.

Let us first introduce a continuous copolymer model. If $B := \{B_t\}_{t \geq 0}$ and $Y := \{Y_t\}_{t \geq 0}$ are two standard Brownian motions. The law of B is denoted by $\widehat{\mathbf{P}}$ and the law of Y is denoted by $\widehat{\mathbb{P}}$. As the notation suggests, B is the continuum analog of S and Y is the continuum analog of the partial sums process associated to ω. The measure of the continuous copolymer is defined by

$$\frac{\mathrm{d}\widehat{\mathbf{P}}_{t,Y}}{\mathrm{d}\widehat{\mathbf{P}}}(B) = \frac{1}{\widehat{Z}_{t,Y}} \exp\left(-2\lambda \int_0^t \mathbf{1}_{B_s \leq 0}\left(\mathrm{d}Y_s + h\,\mathrm{d}s\right)\right), \qquad (6.38)$$

where λ and h are non-negative parameters and $t > 0$. Of course we have

$$\frac{\mathrm{d}\widehat{\mathbf{P}}_{t,Y}}{\mathrm{d}\widehat{\mathbf{P}}}(B) \propto \exp\left(\lambda \int_0^t \mathrm{sign}(B_s)\left(\mathrm{d}Y_s + h\,\mathrm{d}s\right)\right). \qquad (6.39)$$

The free energy is defined in the standard way, however proving the existence of the free energy requires some work.

Proposition 6.6   *The limit of the sequence $\{(1/t)\log \widehat{Z}_{t,Y}\}_t$ exists both $\widehat{\mathbb{P}}(\mathrm{d}Y)$-a.s. and in the $L^1(\widehat{\mathbb{P}}(\mathrm{d}Y))$ sense. Moreover the limit, that we denote by $\widehat{\mathrm{F}}(\lambda, h)$, does not depend on Y.*

We refer to [Giacomin (2004)] for a proof of this result.

The continuous model enjoys the standard properties already verified for the discrete model, notably that

(1) $\widehat{\mathrm{F}}(\lambda, h)$ is non-negative, as one can see by restricting the partition function to the event $\{B : B_s > 0 \text{ for } s \in [1, t]\}$. So one defines $\widehat{\mathcal{L}} := \{(\lambda, h) : \widehat{\mathrm{F}}(\lambda, h) > 0\}$ and $\widehat{\mathcal{D}} := \{(\lambda, h) : \widehat{\mathrm{F}}(\lambda, h) = 0\}$.
(2) by annealing one has that

$$\widehat{\mathrm{F}}(\lambda, h) \leq \lim_{t \to \infty} \frac{1}{t} \log \widehat{\mathbf{E}} \exp\left(\tilde{\beta} \int_0^t \mathbf{1}_{B_s \leq 0}\,\mathrm{d}s\right) = \max\left(\hat{\beta}, 0\right)^2 / 2,$$
$$(6.40)$$

with $\hat{\beta} = 2\lambda(\lambda - h)$. With more work one can also show that $\widehat{\mathrm{F}}(\lambda, 0) \geq c\lambda^2$ for some $c > 0$ (a proof follows by mimicking the proof of Theorem 6.3) and from that one extracts that $\widehat{F}(\lambda, m\lambda) > 0$ for sufficiently small m and λ (but we see below that this implies the result for every λ).

(3) $\widehat{F}(\lambda, h)$ is separately convex and non-decreasing in both variables, which shows in particular that there exists a non-decreasing function $\hat{h}_c : [0, \infty) \to [0, \infty)$ whose graph $\{(\lambda, h) : h = \hat{h}_c(\lambda)\}$ is the phase boundary. The bounds on $\widehat{F}(\cdot, \cdot)$ given in the previous point imply that $h_c(\lambda) > 0$ for every $\lambda > 0$.

With respect to the discrete case however, an important property comes from the (diffusive) scaling property of Brownian motion: for every $\gamma > 0$

$$
\begin{aligned}
\widehat{F}(\lambda, h) &= \lim_{t \to \infty} \frac{1}{t} \widetilde{\mathbb{E}} \log \widehat{\mathbf{E}} \left(\exp \left(-2\lambda \int_0^t \mathbf{1}_{B_s \leq 0} \left(\mathrm{d}(\gamma Y_{s\gamma^{-2}}) + h \, \mathrm{d}s \right) \right) \right) \\
&= \frac{1}{\gamma^2} \lim_{t \to \infty} \frac{\gamma^2}{t} \widetilde{\mathbb{E}} \log \widehat{\mathbf{E}} \left(\exp \left(-2\lambda\gamma \int_0^{t/\gamma^2} \mathbf{1}_{\gamma^{-1}B_{\gamma^2 s} \leq 0} \left(\mathrm{d}Y_s + h\gamma \, \mathrm{d}s \right) \right) \right) \\
&= \gamma^{-2} \widehat{F}(\lambda\gamma, h\gamma). \quad (6.41)
\end{aligned}
$$

Note that we have used the scaling properties for both B and Y. In particular we have $\widehat{F}(\lambda, h) = \widehat{F}(1, h/\lambda)\lambda^2$, which immediately implies that the critical curve is just a straight line, *i.e.* that there exists $\hat{m}_c \geq 0$ such that $\hat{h}_c(\lambda) = \hat{m}_c\lambda$ for every $\lambda > 0$. By the bounds mentioned above we know that $m_c \in (0, 1]$. Another consequence of (6.41) is that $\widehat{F}(\lambda, 0) = \lambda^2 \widehat{F}(1, 0)$ and $\widehat{F}(1, 0) \in (0, 1/2]$. As a matter of fact $\widehat{F}(\lambda, m\lambda) = \lambda^2 \widehat{F}(1, m)$, so the free energy has a quadratic behavior on each of the rays that stem from the origin.

The central result of this section is the following:

Theorem 6.7 *Let S be a $(1, 0)$-walk, that is S is the simple random walk and let ω_1 be either a bounded centered random variable of variance one or a standard Gaussian variable. Then*

$$
\lim_{\gamma \searrow 0} \frac{1}{\gamma^2} F(\gamma\lambda, \gamma h) = \widehat{F}(\lambda, h), \quad (6.42)
$$

for every λ and h.

This result has been proven in [Bolthausen and den Hollander (1997)] for ω_1 taking value ± 1 and it has been generalized in [Giacomin and Toninelli (2005)] by coupling interpolation techniques [Guerra and Toninelli (2002)] and concentration bounds. The importance of such a result is that it suggests (and partly proves) a universal behavior: the weak coupling limit of discrete copolymers is the Brownian copolymer. The universality is with respect to the charge distribution and with respect to the walk (the latter case is an open problem, but it is strongly suggested by Theorem 6.7, see below for more on this).

It is important to remark that Theorem 6.7 implies that $\liminf_{\lambda \searrow 0} h_c(\lambda)/\lambda \geq \hat{m}_c$ ($h_c(\cdot)$ is the critical curve of the random walk model), but not that $\limsup_{\lambda \searrow 0} h_c(\lambda)/\lambda \leq \hat{m}_c$. The statement on the inferior limit is proven by observing that, by the definition of the critical curve, if $m < \hat{m}_c$ then $\widehat{F}(1, m) \geq \varepsilon > 0$ and, by Theorem 6.7, for λ sufficiently small $|(F(\lambda, m\lambda)/\lambda^2) - \widetilde{F}(1, m)| \leq \varepsilon/2$. So $F(\lambda, m\lambda) > 0$ for such values of λ and the claim is proven.

The difficulty in proving the superior limit statement is the same as the one we have encountered in proving the convergence of the critical curve of weakly inhomogeneous models to the one of disordered models, see Figure 4.1. With the (considerable) difference that in this case one can actually show that $\limsup_{\lambda \searrow 0} h_c(\lambda)/\lambda \leq \hat{m}_c$.

Theorem 6.8 *Under the same hypothesis as in Theorem 6.7*

$$\lim_{\lambda \searrow 0} \frac{1}{\lambda} h_c(\lambda) = \hat{m}_c. \tag{6.43}$$

Both Theorem 6.7 and Theorem 6.8 are corollaries of a stronger result proven in [Bolthausen and den Hollander (1997)] for the case in which ω_1 take only value -1 and $+1$. Like for Theorem 6.7, the generalization of Theorem 6.8 to bounded and Gaussian charges is proven in [Giacomin and Toninelli (2005)].

6.3.1 *On more general copolymer models and the Brownian scaling*

The idea of the proof of Theorem 6.7 and of Theorem 6.8 is rather intuitive. It is what is often called a *coarse graining* argument, in the sense that one has to pass on a scale in which one looses some details of the walk and in which only the Brownian universal behavior is left. When the coupling λ is small, the trajectories of the copolymer, at least locally, are modified little by the interaction with the environment. The typical trajectories of the simple random walk are made up of very long excursions: this is of course not specific of the simple random walk, it is just due to the fact that the inter-arrivals of τ, the zero level set, are not integrable. To be more precise, the fraction of $\{1, \ldots, N\}$ covered by finite length returns, say $(1/N) \sum_{j \leq \mathcal{N}_N(\tau)} (\tau_j - \tau_{j-1}) \mathbf{1}_{(\tau_j - \tau_{j-1}) \leq M}$ for any fixed $M > 0$, is $o(1)$ with probability close to 1, as it follows from the Renewal Theorem.

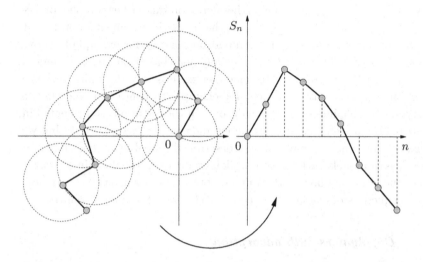

Fig. 6.4   A d-dimensional copolymer model in absence of self-interactions, that is a non-directed walk in a d-dimensional space filled with two solvents (above and below one of the (hyper-)planes of the reference system), may be reduced to a $(1 + 1)$-directed walk model, since the energy depends only on the sign of the component of S perpendicular to the (hyper-)plane. In the figure we draw on the left a walk in a two-dimensional space: the increments are uniformly distributed on a circumference of fixed radius. The walk is then mapped on the right to a directed walk, by projecting on the vertical component. Of course the distribution of the increments of the directed walk is easily computed. We stress that the substantial difference between the resulting model and the one we have treated is that the walk crosses the interface without touching it, so in order to mimic the procedure we have used up to now, one has to introduce a more complex renewal structure, for example the sequence of successive visits to the opposite half-plane and the corresponding *overshoot* variables.

Going back to the copolymer, by the Central Limit Theorem the sum of the charges contributing to the energy inside the long excursions may be approximated by Gaussian random variables, and on the other hand the zero level set of the simple random walk, as well as any renewal with inter-arrival distribution governed by an exponent $\alpha = 1/2$, converges in law to the regenerative set process of Lévy exponent $1/2$, that is the set of zeros of a Brownian motion, so Theorems 6.7 and 6.8 do not appear too surprising. However, transforming such a heuristic argument into a proof demands substantial work and the key idea is a *coarse graining* step, which consists in controlling the approximation of the discrete copolymer with a model in which the energy of the *short* excursions is withdrawn, and then the limit step to Brownian paths is made only on long excursions.

We are not going to give any further detail on this rather complex procedure. But we would like to stress that the heuristic argument we have just outlined strongly suggests that it should be possible for example to prove Theorem 6.7 and Theorem 6.8 for copolymers based on general random walks S with finite variance IID increments. Possibly the finite variance condition has to be strengthened, for example to continuous increments satisfying suitable integrability conditions (in the natural case presented in Figure 6.4 the increments are continuous and compactly supported), but we do expect that the Brownian model describe the weak coupling limit of a large class of models with random walk increments in the domain of attraction of the Gaussian law (with bounded, but otherwise arbitrary, variance) and of sequences of centered independent charges of unitary variance.

6.3.2 *Copolymers with adsorption*

What happens if one superposes a pinning interaction to the copolymer interaction? This amounts to considering the copolymer with adsorption model. We use the notation of (1.66). It should be remarked that if it is easy to show that a sufficiently strong pinning interaction, *i.e.* $\widetilde{h}$ negative and sufficiently large, leads to localization, it is not clear whether a positive $\widetilde{h}$, even very large, always leads to delocalization (we actually know that this is not true for certain values of (λ, h)). Let us see this in some detail: for the purpose of discussing this point we restrict to the case of $\beta = 0$ in (1.66), *i.e.* the pinning interaction is deterministic.

First note that

$$\frac{1}{N} \mathbb{E} \log \mathbf{E} \exp\left(-2\lambda \sum_{n=1}^{N} (\hat{\omega}_n + h)\, \Delta_n - \widetilde{h} \sum_{n=1}^{N} \delta_n\right) =$$
$$\frac{1}{N} \log \mathbf{E} \exp\left(-\widetilde{h} \sum_{n=1}^{N} \delta_n\right) + \frac{1}{N} \mathbb{E} \log \mathbf{E}^{\mathrm{f}}_{N, -\widetilde{h}} \exp\left(-2\lambda \sum_{n=1}^{N} (\hat{\omega}_n + h)\, \Delta_n\right),$$

$$(6.44)$$

where $\mathbf{P}^{\mathrm{f}}_{N, -\widetilde{h}}$ is the homogeneous pinning measure. The limit as $N \to \infty$ of the first term in the right-hand side is precisely the free energy $F(-\tilde{h})$ of the homogeneous pinning model. The second term, by Jensen inequality, the Fubini–Tonelli Theorem and by the fact that $\hat{\omega}$ is a sequence of centered variables, is bounded below by $-2\lambda h$. Therefore $F(-\tilde{h}) > 2\lambda h$ is a sufficient condition for localization, that is verified for $\tilde{h}$ negative and sufficiently large. The question of whether an arbitrarily small pinning $-\tilde{h} > 0$ may lead to localization for $\lambda > 0$ and $h = h_c(\lambda)$ appears to be a very interesting open problem.

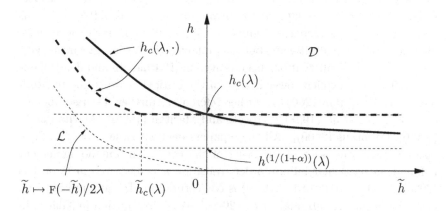

Fig. 6.5 How does $\tilde{h} \mapsto h_c(\lambda, \tilde{h})$ behave? The estimates we know guarantee that it is convex, non-increasing and that it tends to infinity as $\tilde{h}$ tends to $-\infty$ (upper bounds on the growth are easily obtained by annealing). By definition it crosses the vertical axis at $h_c(\lambda)$. By the argument of section 6.2 $h_c(\lambda, \cdot) \geq h^{(1/(1+\alpha))}(\lambda)$. Note that if $h_c(\lambda, \cdot)$ is constant in an interval, then it keeps that value to the right of that interval. Therefore the thick line and the thick dashed line represent, on a qualitative level, the only two options.

Let us turn to the case of $\tilde{h} > 0$. Note that the rare stretch strategy of Section 6.2 is insensitive to the presence of a pinning potential (positive or negative), since for each visit to a rare stretch of length ℓ (which can be made arbitrarily large) it contributes only an energy of size $O(1)$. We can turn this *deficiency* of the argument to our advantage in this case, since it yields that the copolymer is localized whenever $\lambda > 0$ and $h < h^{(1/(1+\alpha))}(\lambda)$, regardless of the value of $\tilde{h}$, a priori a surprising result. Moreover note that the free energy $F(\lambda, h, \tilde{h})$ of the system we are analyzing here is (jointly) convex in h and $\tilde{h}$ and non-increasing in both variables. This directly implies that, for any $\lambda > 0$, the polymer is localized if and only if $h > h_c(\lambda, \tilde{h})$ and $h_c(\lambda, \cdot)$ is convex and non-increasing. Moreover of course $h_c(\lambda, 0) = h_c(\lambda)$, the critical point of the copolymer without adsorption.

We make an attempt of drawing $\widetilde{h} \mapsto h_c(\lambda, \widetilde{h})$ in Figure 6.5 and this should clarify some of the (several) open problems with this apparently simple model.

6.4   Bibliographic Complements

Copolymers have an extremely vast literature. The model we are considering has been first formulated theoretically in [Garel *et al.* (1989)]: in this paper non-rigorous (*replica*) arguments are set forth in favor of a phase diagram that is qualitatively (but also quantitatively) in agreement with Theorem 6.1, a result that has been proven in [Bolthausen and den Hollander (1997)] (the explicit lower bound on the critical curve is due to [Bodineau and Giacomin (2004)]). Other references, further discussed below, are [Stepanow *et al.* (1998)], [Trovato and Maritan (1999)], [Maritan *et al.* (1999)], [Monthus (2000)]. All these papers deal with the $\alpha = 1/2$. They also deal with the issue of the order of the transition, but no agreement is found, in particular infinite order transition is conjectured in [Monthus (2000)], while a lower order (third) is conjectured in [Trovato and Maritan (1999)], see also [Habibzadah *et al.* (2006)]. We point out that in [Giacomin and Toninelli (2006a)] and [Giacomin and Toninelli (2006c)] the case of the copolymer is also addressed: methods and results are extremely close to the ones obtained for the disordered pinning model, *cf.* Section 5.4. Precisely, the result is:

Theorem 6.9   *Consider the copolymer model, with ω satisfying the standard assumptions. If either there exists $C > 0$ such that $\mathbb{P}\left(|\omega_1| > C\right) = 0$ or if ω_1 is a continuous random variable satisfying the entropy condition of Theorem 5.6, then for every $\lambda > 0$ there exists $c(\lambda) > 0$ such that*

$$\mathrm{F}(\lambda, h) \leq (1 + \alpha)c(\lambda)\left(h_c(\lambda) - h\right)^2, \qquad (6.45)$$

for every h.

Therefore the transition is, like in the pinning case, at least of second order.

Complements to Section 6.1

Theorem 6.3 has been proven in [Bolthausen and den Hollander (1997)] in the restricted set-up of simple random walk and ω_1 taking values ± 1. The authors actually mostly aimed to show that the slope of the critical curve at the origin is strictly positive, but in reality they treat the asymmetry as a small perturbation. More in detail, it is straightforward to generalize (6.13) to a lower bound for $F(\lambda, h) + \lambda h$: just replace in the right-hand side $\mathbb{E}\left[\psi(\lambda \omega(0, n])\right]$ with $\mathbb{E}\left[\psi(\lambda \omega(0, n] + hn)\right]$. If we set $q := \inf_n \mathbb{P}(\omega(0, n] \geq 0) > 0$, then

$$\mathbb{E}\left[\psi(\lambda \omega(0, n] + hn)\right] \geq q \mathbb{E}\left[\psi(\lambda \omega(0, n] + hn) \big| \omega(0, n] \geq 0\right], \qquad (6.46)$$

and the last term is bounded below by $q \mathbb{E}\left[\psi(\lambda \omega(0, n]) | \omega(0, n] \geq 0\right]$. Since the argument considers only long excursions and $\psi(\cdot)$ is even, the conditioning at this point will have only a minor effect ($\omega(0, n]$ is almost symmetric and $\omega(0, n] = 0$ is small, for n large) and we are back to the estimate we have made for the case $h = 0$. So the net result is $F(\lambda, h) \geq c\lambda^2 - \lambda h$ for some $c > 0$ and λ small. Therefore $F(\lambda, h) > 0$ is $h < c\lambda$, that is the positivity of the slope.

Complements to Section 6.2

Theorem 6.5 has been proven in [Bodineau and Giacomin (2004)], but the upper bound is a straightforward generalization of what had already been done in [Bolthausen and den Hollander (1997)]. And, like in the pinning case, the constrained annealing procedure of Section 5.3 is of no help [Caravenna and Giacomin (2005)]. The lower bound instead employs a strategy, the rare stretch strategy, that is suggested from the arguments in [Monthus (2000)] where one can find a computation of the critical curve, claimed to be *exact* for the case of $\alpha = 1/2$. There is strong numerical evidence, but yet no proof, that this solution is not correct ([Caravenna *et al.* (2006)], but see also Chapter 9 and [Garel and Monthus (2005c)]). The claim, restricted to Gaussian charges, that the critical curve is a line of slope $2/3$ can be found (also) in [Stepanow *et al.* (1998)] (the argument is based on replica computations). Instead the papers [Garel *et al.* (1989)] and [Trovato and Maritan (1999)], still using replica, claim that the slope is 1. At the moment, *cf.* Section 9.4, numerical computations suggest that this slope lies in between 0.82 and 0.84.

The approach in [Monthus (2000)] deserves a more detailed discussion, both because it has suggested the rigorous proof in Section 6.3 and because it is based on a general renormalization scheme for one-dimensional disordered systems, proposed by D. S. Fisher in [Fisher (1992)], that has enjoyed a sizable success. It has been first applied to quantum Ising chains with transverse field, and then for example to *Random Walks in Random Environments* (RWRE), see *e.g.* [Le Doussal *et al.* (1999)] (for another application, see [Lubensky and D. R. Nelson (2000)]). The results are always claimed to be *exact* or at least *exact close to criticality* and in some case the claim is doubtful. However in the case of RWRE the results are indeed correct: this of course can be assessed because the mathematical understanding of RWRE in one dimension is extremely good [Zeitouni (2004)]. It must be pointed out that the renormalization group approach has been made rigorous in the *slightly* different context of diffusions in random environments [Cheliotis (2005)]. Hence its application to copolymers deserves attention (possibly in order to understand the limitations of the method).

What has been kept in [Bodineau and Giacomin (2004)] of the arguments in [Monthus (2000)] is the strategy of looking for rare stretches and the consequent scales. These scales are then injected by Monthus into Fisher's scheme. It is unclear whether or not, even if these scales are probably not correct, the critical behavior of the copolymer model can be described via Fisher's scheme anyway.

In [Bodineau and Giacomin (2004)] has also been proposed a simplified model, a diluted disordered pinning model, that is supposed to capture some of the features of the mechanism of the copolymer transition. Recent numerical computations however suggest that, according to this model, the slope should be 2/3 and therefore the model is not *good enough*. We believe that such a reduced model is still interesting, both in better understanding the mechanism of transition for the copolymer and for its own sake. In fact it is a very natural and very simple model, but proving the numerical result we have just stated does not seem to be easy.

Complements to Section 6.3

The section is expository and it contains already the references to the results that we state. We point out that, in many of the physics papers on copolymers, the system is informally described as discrete, *i.e.* based on (simple) random walk, but in the end the model is the Brownian copolymer. Whenever numerical computations are performed, the model is again discrete.

Some basic results on the model described in Figure 6.4 can be found in [Caravenna (2005a)].

On copolymers with adsorption there is a certain amount of literature, see *e.g.* [Soteros and Whittington (2004)], but we are not aware of clear answers to the questions we have raised in this section. We point out that a coarse graining result is available also for this model, under suitable conditions, and the limit model is a Brownian copolymer with the energy modified by the appearance of a non-disordered local time term: such a result has been proven in [Pétrélis (2006)] to which we refer for the precise statement.

Chapter 7

The Localized Phase of Disordered Polymers

In this chapter we study the properties of polymers in the localized phase. We focus on extracting properties of the trajectories from the *a priori* much weaker information that the free energy of the system is positive. Moreover such estimates will naturally also yield results on the free energy itself.

In this chapter the charge sequences, ω and $\hat{\omega}$, are considered bi-infinite, *i.e.* they are indexed by $\mathbb{Z}$ rather than $\mathbb{N}$. This is just a technical point and could be avoided, but we adopt it because it simplifies a certain number of expressions.

7.1 A First Tightness Estimate

We start by giving a path localization estimate for copolymers (that is, in the set-up of Chapter 6) and we will stick to the framework of (p, q)-walks. Moreover we will give a *tightness* estimate only for the right endpoint of the polymer (of course in the free endpoint set-up). All these restrictions are motivated by the intent of introducing some of the basic ideas for getting path localization estimates in a transparent way. At the same time this line of presentation follows the historical development of the subject (see Section 7.4). More general models, notably more general return laws, will be treated in the next sections where stronger and more complete results are given. These generalizations involve new ideas, but the key point of the argument that we are presenting now is the starting point, if not the heart, of all localization estimates that we are going to develop. The reader will also notice that, even in the restricted framework we have chosen, the main result is somewhat involved and this highlights some of the intrinsic difficulties in dealing with disordered charge distributions.

We will prove the following:

Theorem 7.1   *For the copolymer model based on a (p,q)-walk (so in particular ω stands for $\hat{\omega}$) if $(\lambda, h) \in \mathcal{L}$, for every $\varepsilon > 0$ there exists a measurable function $C_\varepsilon : \mathbb{R}^\mathbb{Z} \to [0, \infty)$ such that*

$$\mathbf{P}^{\mathsf{f}}_{N,\omega} \left(|S_N| \geq L \right) \leq C_\varepsilon(\theta^N \omega) \exp\left(-(\mathrm{F}(\lambda, h) - \varepsilon) L \right), \qquad (7.1)$$

for every N and every L, $\mathbb{P}(\mathrm{d}\omega)$-a.s..

The *complexity* of this statement lies of course in the pre-factor $C_\varepsilon(\theta^N \omega)$ and that cannot be avoided, as we explain in in Figure 7.1. The following lemma, from which we will extract the proof of Theorem 7.1, should also be of help for the intuition:

Lemma 7.2   *For every N let us consider the transformation $\vartheta_N : \mathbb{R}^\mathbb{Z} \to \mathbb{R}^\mathbb{Z}$ sending $\{\omega_n\}_{n \in \mathbb{Z}}$ to $\{\omega_{N-n+1}\}_{n \in \mathbb{Z}}$ (we have therefore reversed the order of the charges in the polymer chain). With the same set-up as in Theorem 7.1, if $(\lambda, h) \in \mathcal{L}$ for every $\varepsilon > 0$ there exists a measurable function $C'_\varepsilon : \mathbb{R}^\mathbb{Z} \to [0, \infty)$ such that*

$$\mathbf{P}^{\mathsf{f}}_{N,\vartheta_N\omega} \left(|S_N| \geq L \right) \leq C'_\varepsilon(\omega) \exp\left(-(\mathrm{F}(\lambda, h) - \varepsilon) L \right), \qquad (7.2)$$

for every N and every L, $\mathbb{P}(\mathrm{d}\omega)$-a.s..

In the proof we give an explicit expression for $C'_\varepsilon(\omega)$, see (7.7), and it depends only on $\{\omega_n\}_{n \in \mathbb{N}}$. We will see also that $C_\varepsilon(\omega) = C'_\varepsilon(\vartheta_0 \omega)$.

Proof. If $L > N$, then the event $\{|S_N| \geq L\}$ is impossible, so we assume $L \leq N$. On the event $|S_N| \geq L$, we know that the last visit to zero before N is at $N - L$ or before, namely $\tau_{\mathcal{N}_N(\tau)} \leq N - L$. We have

$$\mathbf{P}^{\mathsf{f}}_{N,\vartheta_N\omega} \left(|S_N| \geq L \right) = \sum_{n=L}^{N} Z^{\mathsf{f}}_{N,\vartheta_N\omega} \left(|S_N| \geq L, \tau_{\mathcal{N}_N(\tau)} = N - n \right) / Z^{\mathsf{f}}_{N,\vartheta_N\omega}. \tag{7.3}$$

By the renewal property of τ and the up-down symmetry of the excursions we have

$$Z^{\mathsf{f}}_{N,\vartheta_N\omega}(\tau_{\mathcal{N}_N(\tau)} = N - n) = Z^{\mathsf{c}}_{N-n,\vartheta_N\omega} \frac{1}{2}(1 + \exp(-2\lambda(\omega(0,n] + hn)))\overline{K}(n) \tag{7.4}$$

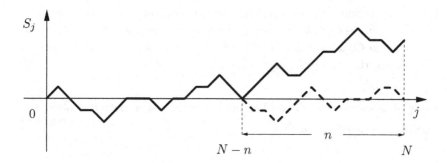

Fig. 7.1 The idea of the proof of Theorem 7.1 is simply to compare the trajectories that after $N - n$ stay positive (or negative) and the dashed trajectories, free to visit the upper and lower half plane as they like. The (strong) Markov property of S or, equivalently, the renewal property of interface visits τ, allows to concentrate our attention on the last portion, of length n of the walk. Observe that varying N the charges in this portion vary and, for typical ω, one observes in the window $(N - n, N]$ arbitrary stretches of charges. In particular, if $\omega_1 \in \{-1, 1\}$, by varying N one can find an arbitrary long stretch of $+1$ ending at N and the endpoint of the polymer long N, even in the localized regime, will appear to be *rather* delocalized. One way of getting around this problem is to attach the environment starting from N backward (this is the composition of a reflection and a translation), *cf.* Lemma 7.2, so that in the window $(N - n, N]$ the charges do not change when N varies. Coming back to the original question, the pre-factor involving the shifting sequence of charges in (7.1) appears.

and this expression is an upper bound on the numerator of the term in the sum in the right-hand side of (7.3). Moreover

$$Z^{\mathrm{f}}_{N,\vartheta_N\omega} \geq Z^{\mathrm{f}}_{N,\vartheta_N\omega}(N - n \in \tau) = Z^{\mathrm{c}}_{N-n,\vartheta_N\omega} Z^{\mathrm{f}}_{n,\vartheta_n\omega} \geq Z^{\mathrm{c}}_{N-n,\vartheta_N\omega} Z^{\mathrm{c}}_{n,\omega}.$$
$$(7.5)$$

In the last step we have exploited the equality $Z^{\mathrm{c}}_{n,\omega} = Z^{\mathrm{c}}_{n,\vartheta_n\omega}$ which is an immediate consequence of the exchangeability of the increments of the sequence τ (or, equivalently, of the reversibility of the random walk paths). Putting (7.3), (7.4) and (7.5) together we have

$$\mathbf{P}^{\mathrm{f}}_{N,\vartheta_N\omega}\left(|S_N| \geq L\right) \leq \sum_{n=L}^{N} \frac{\left(1 + \exp(-2\lambda(\omega(0,n] + hn))\right)\overline{K}(n)}{2Z^{\mathrm{c}}_{n,\omega}}.$$
$$(7.6)$$

We set

$$C'_\varepsilon(\omega) := \sum_{n\in\mathbb{N}} \frac{\left(1 + \exp(-2\lambda(\omega(0,n] + hn))\right)\overline{K}(n)}{2Z^{\mathrm{c}}_{n,\omega}} \exp\left((\mathrm{F}(\lambda, h) - \varepsilon)n\right),$$
$$(7.7)$$

and note that, by the definition of $\mathrm{F}(\lambda, h)$ (Theorem 4.1) and by the Strong

Law of Large Numbers (Theorem A.1, used to control the exceptional values of $\omega(0, n]$), if $\varepsilon \in (0, \mathrm{F}(\lambda, h))$ then $C'_\varepsilon(\omega) < \infty$ $\mathbb{P}(\,d\omega)$-a.s. (so if $C'_\varepsilon(\omega) = \infty$ just redefine $C'_\varepsilon(\omega) = 0$ and $C_\varepsilon(\cdot)$ takes values in $[0, \infty)$). Going back to (7.6) we see that the proof of Lemma 7.2 is complete.   □

Proof of Theorem 7.1. Given Lemma 7.2, it is just a matter of observing that $\vartheta_N^{-1}\omega = \vartheta_0\theta^N\omega(= \theta^{-N}\vartheta_0\omega = \vartheta_N\omega)$.

Theorem 7.1
□

7.2 Quenched versus Quenched Averaged Estimates

We aim at going beyond the results presented in the previous section both in the sense of considering more general models and in the direction of obtaining precise estimates. So we will consider general inter-arrival distributions, even if we treat in detail only the pinning case. Everything that we are going to say goes through for copolymers and for copolymers with adsorption and we refer to Section 7.4 for some details and for references. For the sake of conciseness, we restrict ourselves to the constrained endpoint case: the generalization to the free endpoint case is straightforward.

In what follows the model is the one of Chapter 5 (disordered pinning) and, beyond the standard assumptions on ω (Definition 1.17), we consider only the case in which ω_1 is exponentially integrable (in the sense of Definition 1.16).

We start off by analyzing a quantity whose interest is *a priori* unclear:

Proposition 7.3 *The limit*

$$\lim_{N\to\infty} \frac{1}{N} \log \mathbb{E}\left[1/Z_{N,\omega,\beta,h}^{\mathrm{c}}\right] =: -\mu(\beta, h), \tag{7.8}$$

exists and we have

$$0 \leq \mu(\beta, h) \leq \mathrm{F}(\beta, h), \tag{7.9}$$

for every β and h and, under the additional assumption that ω satisfies a Gaussian concentration inequality (i.e. (A.16) with $\vartheta(t) = c_1 \exp(-c_2 t^2)$), we have

$$\mu(\beta, h) \geq \min\left(\frac{1}{2}\mathrm{F}(\beta, h), \frac{c_2(\mathrm{F}(\beta, h))^2}{32\beta^2}\right). \tag{7.10}$$

Remark 7.4 *Besides showing that $\mu(\beta, h) > 0$ whenever $\mathrm{F}(\beta, h) > 0$, the bound (7.10) is of interest in particular when, for fixed $\beta > 0$, h approaches $h_c(\beta)$ from below (and therefore $\mathrm{F}(\beta, h)$ approaches 0), since it implies that for every $\beta > 0$ there exist two positive constants κ_1 and κ_2 such that*

$$\kappa_1(\mathrm{F}(\beta, h))^2 \leq \mu(\beta, h), \tag{7.11}$$

for $h_c(\beta) - h \leq \kappa_2$.

Proof. By Theorem A.12, the existence of the limit follows from the subadditivity of the sequence $\left\{ \log \mathbb{E} \left[1/Z_{N,\omega,\beta,h}^{\mathrm{c}} \right] \right\}_N$:

$$\mathbb{E}\left[\frac{1}{Z_{N,\omega}^{\mathrm{c}}} \right] \leq \mathbb{E}\left[\frac{1}{Z_{N,\omega}^{\mathrm{c}}(M \in \tau)} \right] = \mathbb{E}\left[\frac{1}{Z_{M,\omega}^{\mathrm{c}} Z_{N-M,\theta^M \omega}^{\mathrm{c}}} \right]$$
$$= \mathbb{E}\left[\frac{1}{Z_{M,\omega}^{\mathrm{c}}} \right] \mathbb{E}\left[\frac{1}{Z_{N-M,\omega}^{\mathrm{c}}} \right], \tag{7.12}$$

for every $M \in \{0, 1, \ldots, N\}$. The standard lower bound $Z_{N,\omega}^{\mathrm{c}} \geq \exp(\beta\omega_N - h)K(N)$ yields the lower bound in (7.9), while the upper bound is an immediate consequence of Jensen's inequality.

In order to establish (7.10) let us observe that we may assume $\mathrm{F}(\beta, h) > 0$, otherwise the inequality we are seeking to prove is the lower bound in (7.9), and let us introduce the event

$$E_N := \left\{ \omega : \frac{1}{2}\mathrm{F}(\beta, h) < \frac{1}{N} \log Z_{N,\omega,\beta,h}^{\mathrm{c}} \right\}. \tag{7.13}$$

We know that there exists N_0 such that $(1/N)\mathbb{E} \log Z_{N,\omega,\beta,h}^{\mathrm{c}} \geq (3/4)\mathrm{F}(\beta, h)$ for $N \geq N_0$. By the concentration property we have that $\mathbb{P}(E_N^{\mathrm{C}})$ is bounded above by $c_1 \exp(-c_2 N(\mathrm{F}(\beta, h)^2/(16\beta^2)))$ (see (4.33)) for such values of N and therefore

$$\mathbb{E}\left[\frac{1}{Z_{N,\omega}^{\mathrm{c}}} \right] \leq \exp\left(-N\mathrm{F}(\beta, h)/2 \right) + \mathbb{E}\left[\frac{1}{Z_{N,\omega,\beta,h}^{\mathrm{c}}}; E_N^{\mathrm{C}} \right], \tag{7.14}$$

and, since by the Cauchy–Schwarz inequality and the concentration estimate we have

$$\mathbb{E}\left[\frac{1}{Z_{N,\omega,\beta,h}^{\mathrm{c}}}; E_N^{\mathrm{C}} \right] \leq \frac{\exp(h)c_1^{1/2}}{K(N)} \mathbb{E}[e^{-2\beta\omega_N}]^{1/2} \exp\left(-\frac{c_2 N(\mathrm{F}(\beta, h))^2}{32\beta^2} \right), \tag{7.15}$$

we obtain (7.10). $\qquad \square$

Given τ and $n \in \mathbb{N}$ we set $gap_n(\tau) := n_R - n_L$, with $n_R = n_R(n, \tau) := \tau_{\mathcal{N}_n(\tau)+1}$ and $n_L = n_L(n, \tau) := \tau_{\mathcal{N}_n(\tau)}$, that is the length of the inter-arrival to which n belongs: of course in the localized phase $gap_n(\tau)$ should be *small*, but the question is how small and how this size depends on β and h. The relevance of $\mu(\beta, h)$ becomes apparent with the following statement that addresses precisely the question:

Theorem 7.5   *Choose $(\beta, h) \in \mathcal{L}$. For every $\varepsilon > 0$*

(1) there exist two measurable functions, $C_\varepsilon^+(\cdot)$ and $C_\varepsilon^-(\cdot)$, from $\mathbb{R}^{\mathbb{Z}}$ to $(0, \infty)$ such that

$$\mathbf{P}_{N,\omega}^{\mathrm{c}}\left(gap_n(\tau) = k\right) \leq C_\varepsilon^+\left(\theta^n \omega\right) \exp\left(-(\mathrm{F}(\beta, h) - \varepsilon)k\right), \qquad (7.16)$$

for every N, every $n \in \{0, 1, \ldots, N-1\}$ and every k, and

$$\mathbf{P}_{N,\omega}^{\mathrm{c}}\left(gap_n(\tau) = k\right) \geq C_\varepsilon^-\left(\theta^n \omega\right) \exp\left(-(\mathrm{F}(\beta, h) + \varepsilon)k\right), \qquad (7.17)$$

for N and n as above and for $k < N/2$.
(2) there exist two positive constants, c_ε^+ and c_ε^-, such that

$$\mathbb{E}\mathbf{P}_{N,\omega}^{\mathrm{c}}\left(gap_n(\tau) = k\right) \leq c_\varepsilon^+ \exp\left(-(\mu(\beta, h) - \varepsilon)k\right), \qquad (7.18)$$

for every N, every $n \in \{0, 1, \ldots, N-1\}$ and every k, and

$$\mathbb{E}\mathbf{P}_{N,\omega}^{\mathrm{c}}\left(gap_n(\tau) = k\right) \geq c_\varepsilon^- \exp\left(-(\mu(\beta, h) + \varepsilon)k\right), \qquad (7.19)$$

for N and n as above and for $k < N/2$.

We split the proof in upper and lower bounds.
Proof of Theorem 7.5, upper bounds. The argument generalizes the argument in the proof of Theorem 7.1. The event $\{gap_n(\tau) = k\}$ may be written as the disjoint union of the events $E_{n_-,n_+} := \{n_L(n,\tau) = n_-, n_R(n,\tau) = n_+\}$ with $0 \leq n_- \leq n$, $n < n_+ \leq N$ and $n_+ - n_- = k$. Note that this disjoint union is over at most k events. We have

$$\mathbf{P}_{N,\omega}^{\mathrm{c}}\left(E_{n_-,n_+}\right) \leq \frac{Z_{N,\omega}^{\mathrm{c}}\left(E_{n_-,n_+}\right)}{Z_{N,\omega}^{\mathrm{c}}\left(\{n_-, n_+\} \in \tau\right)} = \frac{K(n_+ - n_-)\exp(\beta\omega_{n_+} - h)}{Z_{n_+ - n_-,\theta^{n_-}\omega}^{\mathrm{c}}}, \qquad (7.20)$$

where, in the last step, we have used the renewal property. Note that

$$\mathbb{E}\mathbf{P}_{N,\omega}^{\mathrm{c}}\left(E_{n_-,n_+}\right) \leq K(k)\mathbb{E}\left[\frac{\exp(\beta\omega_k - h)}{Z_{k,\omega}^{\mathrm{c}}}\right] \overset{k\to\infty}{\asymp} \exp\left(-k\mu(\beta, h)\right), \qquad (7.21)$$

as it follows from the definition of $\mu(\beta, h)$, since the expectation in the middle term of (7.21) is equal to $\mathbb{E}\left[1/Z_{k,\omega}^c\right]$ divided by $\mathbb{E}\left[\exp(-\beta\omega_k + h)\right]$ (a constant). The bound (7.18) is therefore proven since summing over n_- and n_+ introduces at most a factor k.

For the quenched estimate we go back to (7.20) and we observe that it suffices to show that

$$\liminf_{k\to\infty} \inf_{\substack{n_-,n_+:\, n_+-n_-=k \\ n_-\leq 0, n_+>0}} \frac{1}{k} \log Z_{k,\theta^{-n-}\omega}^c \geq \text{F}(\beta, h), \qquad (7.22)$$

because the term in the numerator is easily controlled by using the fact that $\lim_{N\to\infty} \omega_N/N = 0$, $\mathbb{P}(\mathrm{d}\omega)$-a.s., by integrability of ω_1 and independence of the sequence ω (see Appendix A.1). For what concerns (7.22) it suffices to observe that

$$\frac{1}{k} \log Z_{k,\theta^{-n}-\omega}^c \geq$$
$$\frac{|n_-|}{k}\left(\frac{1}{|n_-|} \log Z_{|n_-|,R\omega}^c\right) + \frac{n_+}{k}\left(\frac{1}{n_+} \log Z_{n_+,\omega}^c\right) + \frac{\beta\omega_0 - h}{k}, \qquad (7.23)$$

with $(R\omega)_n = \omega_{-n}$ and if $n_- = 0$ the right-hand side contains only the second term. But since, by Theorem 4.1, for every $\varepsilon > 0$, $\mathbb{P}(\mathrm{d}\omega)$-a.s. there exists $n_0 = n_0(\omega) \in \mathbb{N}$ such that both $(1/n)\log Z_{n,\omega}^c$ and $(1/n)\log Z_{n,R\omega}^c$ are bounded below by $\text{F}(\beta, h) - \varepsilon/2$ for $n \geq n_0$, and at the same time $|\beta\omega_0 - h|/k \leq \varepsilon/2$, we obtain (7.22). $\qquad\square$

Proof of Theorem 7.5, lower bounds. We exploit the fact that for $n+k \leq N$

$$\mathbf{P}_{N,\omega}^c\left(gap_n(\tau) = k\right) \geq \mathbf{P}_{N,\omega}^c\left(E_{n,n+k}\right) = \frac{Z_{N,\omega}^c\left(E_{n,n+k}\right)}{Z_{N,\omega}^c}. \qquad (7.24)$$

In order for k to range up to $N/2$ we assume $n \leq N/2$. If $n > N/2$ one should simply repeat the argument by using the event $E_{n-k+1,n+1}$. The technical difficulty comes from the evaluation of the denominator. The following lemma is of help:

Lemma 7.6 *There exist two positive constants c and κ such that for every N, every $j \in \{1,\ldots N - 1\}$ and every ω we have*

$$Z_{N,\omega}^c\left(j \in \tau, \Omega_S^j\right) \geq \frac{c}{\zeta(\omega_j)\left(\min(j, N - j)\right)^\kappa} Z_{N,\omega}^c\left(\Omega_S^j\right), \qquad (7.25)$$

where $\zeta(r) := \exp(|\beta r - h|)$ and Ω_S^j is in the σ-algebra generated by $\tau \cap [1, j-1]$.

We postpone the proof of the lemma and continue with the main argument. At this point one might be tempted to apply the lemma directly to the denominator in the rightmost term of (7.24) and replace $Z_{N,\omega}^c$ with $Z_{N,\omega}^c(\{n, n+k\} \in \tau)$. However the (multiplicative) error term would contain a power of n and not only a power of k. In order to get around this difficulty we exploit the fact that, since we are in the localized regime, with high probability there is a renewal to the left of n, at a distance at most (say) k^2. Namely we introduce the event $\Omega_S(n, k) := \{\tau \cap [n-k^2, n-1] \neq \emptyset\}$ and we observe that we can write

$$\frac{1}{Z_{N,\omega}^c} = \frac{1}{Z_{N,\omega}^c(\Omega_S(n,k))} \left(1 - \mathbf{P}_{N,\omega}^c\left(\Omega_S(n,k)^c\right)\right), \qquad (7.26)$$

and if we introduce $T_n := \max(\tau \cap [0, n-1])$, that is the first renewal to the left of n, then

$$Z_{N,\omega}^c\left(\Omega_S(n,k)\right) = \sum_{\substack{j=1,\ldots,k^2 \\ n-j \geq 0}} Z_{N,\omega}^c\left(\Omega_S(n,k),\, T_n = n-j\right)$$

$$= \sum_{\substack{j=1,\ldots,k^2 \\ n-j \geq 0}} Z_{n-j,\omega}^c Z_{N-(n-j),\theta^{n-j}\omega}^c\left(\tau \cap \{1,\ldots,j-1\} = \emptyset\right).$$
$$(7.27)$$

Now let us apply Lemma 7.6 twice, the first time with j replaced by $j+k$ and $\Omega_S^{j+k} = \{\tau \cap [1, j-1] = \emptyset\}$ and the second time with $\Omega_S^j = \{\tau \cap [1, j-1] = \emptyset\}$, to obtain

$$Z_{M,\omega}^c\left(\{j, j+k\} \in \tau,\, \tau \cap \{1,\ldots,j-1\} = \emptyset\right)$$
$$\geq \frac{c^2 Z_{M,\omega}^c\left(\tau \cap \{1,\ldots,j-1\} = \emptyset\right)}{\zeta(\omega_j)\zeta(\omega_{j+k})\left(\min(j, M-j)\min(j+k, M-j-k)\right)^\kappa}, \qquad (7.28)$$

and, given the renewals in j and $j+k$, the left-hand side factors into the product of three terms. Plugging this estimate, with $M = N - (n-j)$ and ω replaced by $\theta^{n-j}\omega$, into (7.27) and resumming over j we get to

$$Z_{N,\omega}^c\left(\Omega_S(n,k)\right) \leq ck^{4\kappa}\zeta(\omega_n)\zeta(\omega_{n+k})Z_{n,\omega}^c Z_{k,\theta^n\omega}^c Z_{N-n-k,\theta^{n+k}\omega}^c, \qquad (7.29)$$

for some positive constant c.

Since by the renewal property

$$Z_{N,\omega}^{c}(E_{n,n+k}) = Z_{n,\omega}^{c}K(k)\exp(\beta\omega_{n+k} - h)Z_{N-n-k,\theta^{n+k}\omega}^{c}, \qquad (7.30)$$

putting (7.24), (7.26) and (7.29) together we get to

$$\mathbf{P}_{N,\omega}^{c}(gap_n(\tau) = k)$$

$$\geq \frac{1}{ck^{4\kappa}\zeta(\omega_n)\zeta(\omega_{n+k})}\left[\frac{K(k)e^{\beta\omega_{n+k}-h}}{Z_{k,\theta^n\omega}^{c}}\right]\left(1 - \mathbf{P}_{N,\omega}^{c}\left(\Omega_S(n,k)^{c}\right)\right)$$

$$\geq \frac{1}{ck^{4\kappa}\zeta(\omega_n)\zeta(\omega_{n+k})}\left(\frac{K(k)e^{\beta\omega_{n+k}-h}}{Z_{k,\theta^n\omega}^{c}} - \mathbf{P}_{N,\omega}^{c}\left(\Omega_S(n,k)^{c}\right)\right), \qquad (7.31)$$

where in the second step we have used the fact that the term between brackets is equal to $\mathbf{P}_{k,\theta^n\omega}^{c}(E_{n,n+k})$ and hence it is smaller than 1.

The bound (7.17) is now obtained by using Theorem 4.1, which guarantees that $Z_{k,\omega}^{c}\exp(-(\mathrm{F}(\beta,h)+\varepsilon)k)$ tends to 0 $\mathbb{P}(\,d\omega)$-a.s. as k tends to infinity, as well as $\mathbf{P}_{N,\omega}^{c}\left(\Omega_S(n,k)^{c}\right)\exp(+ck^2)$ for some $c > 0$. The latter estimate follows by applying (7.16)), since if $\tau \in \Omega_S(n,k)^{c}$ then n (to be precise: $n - 1$) belongs to a *gap* wide at least k^2 steps. Of course one uses also that, $\mathbb{P}(\,d\omega)$-a.s., $\omega_k/k \xrightarrow{k\to\infty} 0$.

In order to obtain the bound (7.19) we go back to (7.31) and we have to estimate two terms. The first one is controlled by by applying Proposition 7.3, as it has been done in the last part of the proof of (7.18): this yields the leading part of the estimate. Fo the second term we use the Cauchy–Schwarz inequality to get

$$\mathbb{E}\left[\frac{1}{\zeta(\omega_n)\zeta(\omega_{n+k})}\mathbf{P}_{N,\omega}^{c}\left(\Omega_S(n,k)^{c}\right)\right] \leq$$

$$\mathbb{E}\left[\frac{1}{(\zeta(\omega_n)\zeta(\omega_{n+k}))^2}\right]^{1/2}\mathbb{E}\left[\mathbf{P}_{N,\omega}^{c}\left(\Omega_S(n,k)^{c}\right)\right]^{1/2}. \qquad (7.32)$$

At this point it suffices to observe that the second factor in the last term is bounded above by $\exp(-ck^2)$ for some $c > 0$ (in this case we can choose $c = \mu(\beta,h)/2$). This follows by applying (7.18) after having decomposed the event $\Omega_S(n,k)^{c}$ according to the location of the endpoints of the excursion that contains $\{n - k^2,\ldots,n-1\}$. $\qquad\Box$

Proof of Lemma 7.6. It is based on a trick already used in Chapter 4. Note that it suffices to take $\Omega_S^{j} = \{\tau\cap[1,j-1] = B\}$, where $B \subset \{1,\ldots,j-1\}$.

We first write

$$Z^{c}_{N,\omega}\left(\Omega^j_S\right) = Z^{c}_{N,\omega}\left(j \in \tau,\, \Omega^j_S\right)$$

$$+ \sum_{j_+=j+1}^{N} Z^{c}_{j_-,\omega}\left(\tau \cap [1,j_-] = B \cap [1,j_-]\right)K(j_+ - j_-)e^{\beta\omega_{j_+}-h}Z^{c}_{N-j_+,\theta^{j}+\omega},$$

$$(7.33)$$

where $j_- = \max B \cup \{0\}$. Then we note that, by Definition 1.4 and the properties of slowly varying functions, there exists $c > 0$ such that

$$\frac{K(j_+ - j_-)}{K(j - j_-)K(j_+ - j)} \le c\min\,(j - j_-, j_+ - j)^{\kappa} \le c\min\,(j, N - j)^{\kappa},$$

$$(7.34)$$

where $\kappa := 2(1+\alpha)$. Inserting the latter estimate into (7.33) and performing the sum one obtains that

$$c\min\,(j, N - j)^{\kappa}\exp\left(|\beta\omega_j - h|\right)Z^{c}_{N,\omega}\left(j \in \tau, \Omega^j_S\right), \qquad (7.35)$$

bounds from above the second line in (7.33) and the proof is easily completed. $\hfill$ Theorem 7.6 $\square$

7.3 Quenched Averaged Estimates and the Infinite Volume Polymer Measure

In this section we will consider functions real valued functions $f(\cdot)$ defined on the power set of $\mathbb{N} \cup \{0\}$. Of course such a function may be viewed as an application from $\{0,1\}^{\mathbb{N}\cup\{0\}}$, equipped with the product topology, to $\mathbb{R}$. As it is standard, we say that $f(\cdot)$ is local if there exists $n \in \mathbb{N}$ such that $f(x) = f(y)$ whenever $x_j = y_j$ for $j = 0, 1, \ldots, n$. In terms of renewals this means that $f(\tau) = f(\tau')$ if $\tau \cap [0,n] = \tau' \cap [0,n]$. We denote by $\mathcal{S}(f)$ the *support* of f, that is the intersection of the sets $E \subset \mathbb{N} \cup \{0\}$ with the property that $f(x) = f(y)$ whenever $x_j = y_j$ for every $j \in E$.

The main result of this section is self-explanatory:

Theorem 7.7 *For every $(\beta, h) \in \mathcal{L}$ there exist two positive constants c_1*

and c_2 such that if f and g are bounded local functions we have

$$\mathbb{E}\left[\left|\mathbf{E}^{c}_{N,\omega}(f(\tau)g(\tau)) - \mathbf{E}^{c}_{N,\omega}(f(\tau))\mathbf{E}^{c}_{N,\omega}(g(\tau))\right|\right]$$
$$\leq c_1\|f\|_\infty\|g\|_\infty \exp\left(-c_2 d(\mathcal{S}(f),\mathcal{S}(g))\right), \quad (7.36)$$

with $d(I,J) := \min\{|i-j| : i \in I, j \in J\}$, I and $J \subset \mathbb{N}\cup\{0\}$ $(d(I,J) := \infty$ if either I or J are empty), and if $\sup \mathcal{S}(f) \leq k$ then

$$\sup_{N>k} \mathbb{E}\left[\left|\mathbf{E}^{c}_{N,\omega}(f(\tau)) - \mathbf{E}^{c}_{k,\omega}(f(\tau))\right|\right] \leq c_1\|f\|_\infty \exp\left(-c_2 d(\mathcal{S}(f),\{k\})\right).$$
$$(7.37)$$

The most important consequence of Theorem 7.7 is the following:

Corollary 7.8 *The limit of the sequence $\left\{\mathbf{P}^{c}_{N,\omega}\right\}_N$ of probability measures exists $\mathbb{P}(d\omega)$–almost surely.*

Proof. It suffices to show that $\left\{\mathbf{E}^{c}_{N,\omega}[f(\tau)]\right\}_N$ converges $\mathbb{P}(d\omega)$–a.s. for every local function. Note that (7.37) implies that $\left\{\mathbf{E}^{c}_{N,\omega}[f(\tau)]\right\}_N$ is a Cauchy sequence in $L^1(\mathbb{P})$, and therefore it converges (in $L^1(\mathbb{P})$) to a limit random variable that we denote by $\mathbf{E}^{c}_{\infty,\omega}[f(\tau)]$. We can therefore consider the limit as $N \to \infty$ in (7.37) itself and by the Fubini–Tonelli Theorem we obtain

$$\mathbb{E}\sum_{k}\left|\mathbf{E}^{c}_{\infty,\omega}[f(\tau)] - \mathbf{E}^{c}_{k,\omega}[f(\tau)]\right| < \infty, \quad (7.38)$$

so that the series inside the brackets is $\mathbb{P}(d\omega)$–a.s. convergent and the terms of the series tend to zero $\mathbb{P}(d\omega)$–a.s.. $\qquad\square$

It is interesting to note that from (7.36) one can extract an almost sure statement on the decay of correlations for the infinite volume measure $\mathbf{P}^{c}_{\infty,\omega}$:

Corollary 7.9 *For $(\beta,h) \in \mathcal{L}$ there exists $c > 0$ such that for every (local and bounded) f and g we can find a random variable $C_{f,g}(\omega)$, $C_{f,g}(\omega) < \infty$ $\mathbb{P}(d\omega)$–a.s., such that*

$$\left|\mathbf{E}^{c}_{\infty,\omega}\left[f(\tau)g(\tau+k)\right] - \mathbf{E}^{c}_{\infty,\omega}\left[f(\tau)\right]\mathbf{E}^{c}_{\infty,\omega}\left[g(\tau+k)\right]\right| \leq C_{f,g}(\omega)\exp(-ck),$$
$$(7.39)$$

for every $k \in \mathbb{N}$.

Proof. Take the limit $N \to \infty$ in (7.36) to obtain that there exists $k_0 \in \mathbb{N}$ such that

$$\mathbb{E}\left[\exp(c_2 k/2) \left| \mathbf{E}_{\infty,\omega}^{c} \left[f(\tau)g(\tau + k)\right] - \mathbf{E}_{\infty,\omega}^{c} \left[f(\tau)\right] \mathbf{E}_{\infty,\omega}^{c} \left[g(\tau + k)\right] \right| \right] \\ \leq \exp(-c_2 k/4), \quad (7.40)$$

for $k \geq k_0$. Therefore, by summing over k and by applying the Fubini–Tonelli Theorem, we obtain that

$$\mathbb{E}\left[\sum_{k} \exp(c_2 k/2) \left| \mathbf{E}_{\infty,\omega}^{c} \left[f(\tau)g(\tau + k)\right] - \mathbf{E}_{\infty,\omega}^{c} \left[f(\tau)\right] \mathbf{E}_{\infty,\omega}^{c} \left[g(\tau + k)\right] \right| \right],$$
$$(7.41)$$

is finite, which implies the claim. $\square$

In the proof of Theorem 7.7 we will make use of the following lemma, which is proven in Appendix B.3. It simply states that on a large interval the contact sets of two independent copies of the localized systems meet in at least one point with large probability, if one averages over the disorder.

Lemma 7.10 *Let $\tau^{(1)}$ and $\tau^{(2)}$ be two independent copies of τ. For every $(\beta, h) \in \mathcal{L}$ there exist two positive constants c_1 and c_2 such that for every N, every u and l, with $0 < u < l < N$, we have*

$$\mathbb{E}\left[\mathbf{P}_{N,\omega}^{c} \otimes \mathbf{P}_{N,\omega}^{c} \left(\tau^{(1)} \cap \tau^{(2)} \cap \{u, \ldots, l\} = \emptyset\right)\right] \leq c_1 \exp(-c_2(l - u)). \tag{7.42}$$

Proof of Theorem 7.7. Set $u := 1 + \max \mathcal{S}(f)$ and $l := -1 + \min \mathcal{S}(g)$. We now assume $u < l$, since otherwise the statement is trivial (choose $c_1 \geq 2$), and we set $E_{u,l} := \{\tau^{(1)} \cap \tau^{(2)} \cap [u, l] = \emptyset\}$, with $\tau^{(1)}$ and $\tau^{(2)}$ two independent copies of τ (below $\mathbf{P}_{N,\omega}^{\otimes 2}$ stands for $\mathbf{P}_{N,\omega}^{c} \otimes \mathbf{P}_{N,\omega}^{c}$). We write

$$\mathbf{E}_{N,\omega}^{c}(f(\tau)g(\tau)) - \mathbf{E}_{N,\omega}^{c}(f(\tau))\mathbf{E}_{N,\omega}^{c}(g(\tau)) = \\ \mathbf{E}_{N,\omega}^{\otimes 2}\left[f(\tau^{(1)})g(\tau^{(1)}) - f(\tau^{(1)})g(\tau^{(2)})\right], \quad (7.43)$$

and observe that, using the random variable $T_u := \inf\{\tau^{(1)} \cap \tau^{(2)} \cap [u, \infty)\}$, we have

$$\mathbf{E}_{N,\omega}^{\otimes 2} \left[f(\tau^{(1)}) \left(g(\tau^{(1)}) - g(\tau^{(2)}) \right) ; E_{u,l}^{\complement} \right] =$$
$$\sum_{n=u}^{l} \mathbf{E}_{N,\omega}^{\otimes 2} \left[f(\tau^{(1)}) \left(g(\tau^{(1)}) - g(\tau^{(2)}) \right) ; T_u = n \right], \quad (7.44)$$

and by applying the renewal property, recalling that $\mathcal{S}(f) < u$, one sees that the terms in the sum are equal to

$$\mathbf{E}_{N,\omega}^{\otimes 2} \left[f(\tau^{(1)}); T_u = n \right] \mathbf{E}_{N,\omega}^{\otimes 2} \left[g(\tau^{(1)}) - g(\tau^{(2)}) \middle| T_u = n \right]. \quad (7.45)$$

But, by symmetry, $\mathbf{E}_{N,\omega} \left[g(\tau^{(1)}) \middle| T_u = n \right] = \mathbf{E}_{N,\omega} \left[g(\tau^{(2)}) \middle| T_u = n \right]$ and we conclude that

$$\mathbf{E}_{N,\omega}^{\otimes 2} \left[f(\tau^{(1)}) g(\tau^{(1)}) - f(\tau^{(1)}) g(\tau^{(2)}); E_{u,l}^{\complement} \right] = 0, \quad (7.46)$$

and therefore (7.43) holds also if in the right-hand side we insert (the indicator function of) the event $E_{u,l}$. The probability of $E_{u,l}$ is estimated in Lemma 7.10 and this yields (7.36).

In order to establish (7.37) we observe that, since $\max \mathcal{S}(f) < k$, the following identity holds:

$$\mathbf{E}_{k,\omega}^{\complement} \left[f(\tau) \right] = \frac{\mathbf{E}_{N,\omega}^{\complement} \left[f(\tau) \delta_k \right]}{\mathbf{E}_{N,\omega}^{\complement} \left[\delta_k \right]}. \quad (7.47)$$

By (7.36), we have

$$\mathbf{E} \left[\left| \mathbf{E}_{N,\omega}^{\complement} \left[f(\tau) \delta_k \right] - \mathbf{E}_{N,\omega}^{\complement} \left[f(\tau) \right] \mathbf{E}_{N,\omega}^{\complement} \left[\delta_k \right] \right|^2 \right] \leq 2 c_1 \| f \|_{\infty}^2 \exp \left(-c_2 k / 2 \right), \quad (7.48)$$

for $k > 2 \max \mathcal{S}(f)$: we have used the fact that the term in the absolute value in the left-hand side is bounded by $2 \| f \|_{\infty}$, for every ω. On the other hand, by Lemma 7.6, we have $\mathbf{E}_{N,\omega}^{\complement} [\delta_k] = Z_{N,\omega}^{\complement} (k \in \tau) / Z_{N,\omega}^{\complement} \geq 1 / (c \zeta(\omega_k) k^{\kappa})$, so that

$$\mathbb{E}[1 / \mathbf{E}_{N,\omega}^{\complement} [\delta_k]^2] \leq c' k^{2\kappa}, \quad (7.49)$$

for some positive constant c' and for every $k < N$. Now consider the left-hand side of (7.37): use (7.47) and apply the Cauchy–Schwarz inequality, along with (7.48) and (7.49), to conclude.

Theorem 7.10

$\square$

7.4 Bibliographic Complements

Complements to Section 7.1

This section is adapted from [Sinai (1993)], in which a copolymer based on simple random walk and with symmetric charges taking value ± 1 ($h = 0$) is considered. The aim of [Sinai (1993)] is to prove localization for every $\lambda > 0$ and this is achieved directly by a path estimate, without proving that the free energy is positive (in fact, without using the free energy at all). In particular therefore Sinai's result contains an estimate that is much rougher than the one given in Theorem 7.1, even if the idea can be directly upgraded (and that is what we did). It is natural to ask whether the result in Theorem 7.1 is optimal and the answer is no: we have estimated well, see below, the probability of having a deviation in the last point of the contact set, but we have treated the path of the random walk *very roughly* in the last excursion. Taking into account the probability that in the last excursion the walk may have a less atypical behavior than the one we have used (ballistic) leads to a variational problem (this has been done in [Albeverio and Zhou (1996)] for the $h = 0$ case). A version of Theorem 7.1 can also be found in [Biskup and den Hollander (1999)].

Complements to Section 7.2

The results of this section are taken from [Giacomin and Toninelli (2006b)], where the general case of copolymers with adsorption is treated. The exponent μ first appears in [Albeverio and Zhou (1996)], where the case of copolymers, with $h = 0$, $\omega_1 = \pm 1$ and no adsorption, is considered. The result proven in [Albeverio and Zhou (1996)] is that in the localized phase for every $\varepsilon > 0$

$$\lim_{N \to \infty} \mathbf{P}^{\mathrm{f}}_{N,\omega} \left(\sup_{n \in \{0,\dots,N\}} gap_n(\tau) \in (\mu - \varepsilon, \mu + \varepsilon) \log N \right) = 1, \qquad (7.50)$$

in $\mathbb{P}$–probability. In [Giacomin and Toninelli (2006b)] this result is generalized (and the proof is different) and it is shown, under suitable conditions on the law of ω_1, that $\mu < \mathrm{F}$. The latter inequality is therefore telling us also that the largest typical excursion, sharply estimated by (7.50), are due to atypical stretches in the environment, since by Theorem 7.5 typical large excursions are controlled by F.

It would be extremely interesting to prove (if it is true) that F and

μ vanish in a different way approaching criticality. Related to this issue, but not resolving it, we point out that in [Toninelli (2006)] a bound that sensibly improves on (7.10) is proven.

Complements to Section 7.3

It is natural to ask what the *optimal* value for c_2, in Theorem 7.7, is. Lemma 7.10 is rough (no effort is made to track the constants, but the method itself does not appear to be very sharp). Based on Theorem 7.5(2), one is possibly tempted to say that c_2 can be chosen arbitrarily close to $\mu(\beta, h)$, but this is not completely evident. In the restricted set-up of the disordered pinning model based on simple random walk with hard wall condition, with law denoted by $\mathbf{P}_{N,\omega}^+$, however a precise result can be proven [Toninelli (2006)]. Precisely the result is that for every $(\beta, h) \in \mathcal{L}$ the limit

$$- \lim_{k \to \infty} \frac{1}{k} \log \mathbb{E} \left[\mathbf{E}_{\infty,\omega}^+ (\delta_\ell \delta_{\ell+k}) - \mathbf{E}_{\infty,\omega}^+ (\delta_\ell) \mathbf{E}_{\infty,\omega}^+ (\delta_{\ell+k}) \right] = \mu(\beta, h), \quad (7.51)$$

where $\mathbf{P}_{\infty,\omega}^+$ is the limit of $\left\{ \mathbf{P}_{N,\omega}^+ \right\}_N$ and k, ℓ and k are even and $\delta_n = 1_{S_n=0}$. This precise result however does not seem to extend easily to more general renewal processes: the renewal on which $\mathbf{P}_{N,\omega}^+$ is based is very particular because it is the zero level set of a Markov process with increments ± 1. One can then couple two copies of this process in a very efficient way, see [Toninelli (2006)]. This point is related to the same issue in the non-disordered case and with the problem of finding the speed of convergence to equilibrium for renewal processes, see Section 2.5.

In [Toninelli (2006)] one can find also the quenched result corresponding to (7.51). It will come as no surprise, given for example Theorem 7.5, that for the quenched statement one has to replace $\mu(\beta, h)$ with $\mathrm{F}(\beta, h)$. However, again, things are not so straightforward and, in particular, the proof is restricted to the case of disordered pinning with hard wall constraint.

As Theorem 7.7, Corollary 7.8 has been proven in [Giacomin and Toninelli (2006b)]. However an analogous result, for copolymers based on simple random walks and ω_1 taking values ± 1 is proven in [Biskup and den Hollander (1999)], where a *Gibbsian viewpoint* is taken [Georgii (1988)]. Both the semi-infinite set-up, like ours, and the doubly-infinite set-up is considered (for the localized phase). In particular one can find there results like that the limit measure is *covariant* by translations (it cannot be invariant under translation, but it becomes invariant if we translate also ω at the same time) and uniqueness results in the class of measures with sub-linear growth at infinity.

In [Giacomin and Toninelli (2006b)] it is also shown, in the general context of copolymers with adsorption, that the free energy is C^∞ in $\mathcal{L}$. This is actually a consequence of Theorem 7.7 (rather, of the general version of this theorem), coupled with the procedure and the estimates in [von Dreifus *et al.* (1995)], the idea being that taking the derivative(s) of $(1/N) \log Z^c_{N,\omega}$ with respect to the parameters $\underline{v}$ one obtains terms that are correlation functions and Theorem 7.7 provides the bound that allow passing to the limit $N \to \infty$.

Chapter 8

The Delocalized Phase of Disordered Polymers

In the most general context and with the notation of (1.66), if $\underline{v} \in \mathring{\mathcal{D}}$, by convexity we have that

$$0 = \partial_h \mathrm{F}(\underline{v}) = -2\lambda \lim_{N \to \infty} \frac{1}{N} \mathbf{E}_{N,\omega,\underline{v}} \left[\sum_{n=1}^{N} \Delta_n \right], \tag{8.1}$$

as well as

$$0 = \partial_{\tilde{h}} \mathrm{F}(\underline{v}) = - \lim_{N \to \infty} \frac{1}{N} \mathbf{E}_{N,\omega,\underline{v}} \left[\sum_{n=1}^{N} \delta_n \right], \tag{8.2}$$

$\mathbb{P}(\,\mathrm{d}\omega)$–a.s.. Of course if Δ_n is defined to be equal to 1 when S_n is at the interface, the second statement is (indirectly) included in the first. We will see below that it is included with any definition of the sign of the monomer at the interface, *cf.* Proposition 8.1. Instead if $\lambda = 0$, that is if there is no copolymer interaction, then one has only the second statement and in fact there is no reason for the polymer to have a preference for one side of the interface or the other.

The result (8.1), respectively (8.2), extends to the whole of $\mathcal{D}$ if the free energy is C^1 in h, respectively in $\tilde{h}$, at the transition (see Appendix A.1.1).

The statements (8.1) and (8.2) are clear statements of delocalization and in this sense they are satisfactory. However what we have obtained for homogeneous and weakly inhomogeneous systems suggests that much stronger statements hold. Notice in fact that the statements above are just saying that the number of the monomers of a delocalized polymer that are at the interface or in the unfavorable solvent is $o(N)$, while for homogeneous and weakly inhomogeneous systems one can push $o(N)$ down to $O(1)$. The gap is too evident and in this chapter we will give some results and techniques going toward a better understanding of the delocalized phase, that, however, remains rather elusive.

8.1   Large Deviations and Delocalization

Beyond what is suggested by homogeneous systems, the entropic repulsion phenomenon, *cf.* Section 1.3, should play a role. If it is not favorable for the polymer to visit a certain portion of space, then the polymer possibly chooses to keep very far from such a region. Neither (8.1) nor (8.2) seem to contain such an information.

The aim of the next result addresses this issue and it contains an improvement of (8.1) and (8.2). However it is a direct consequence of Large Deviation estimates and it should be read as pushing the boundary of what one can do with Large Deviations. Precisely the way in which the delocalized regime is defined, *i.e.* the regime in which the energetic contribution is $o(N)$, is strongly hinting that with Large Deviations alone we cannot go too far. This is also suggested by the precise (polynomial!) estimates we have used to study the trajectories of homogeneous polymers in the delocalized regime.

The result we are presenting now deals with (p, q)-walks, later on in the chapter we will work in a more general context. In this section the results hold indifferently for $a = \mathsf{c}$ and $a = \mathsf{f}$.

Proposition 8.1   *Fix an arbitrary $L > 0$ and $\underline{v} \in \overset{\circ}{\mathcal{D}}$. If $\lambda > 0$ then*

$$\lim_{N \to \infty} \frac{1}{N} \mathbf{E}_{N,\omega,\underline{v}}^{a} \left[\sum_{n=1}^{N} 1_{S_n > L} \right] = 1, \tag{8.3}$$

$\mathbb{P}(\, \mathrm{d}\omega)$*-a.s.. If instead $\lambda = 0$, then*

$$\lim_{N \to \infty} \frac{1}{N} \mathbf{E}_{N,\omega,\underline{v}}^{a} \left[\sum_{n=1}^{N} 1_{|S_n| > L} \right] = 1, \tag{8.4}$$

$\mathbb{P}(\, \mathrm{d}\omega)$*-a.s..*

The statement has two parts because if $h = 0$ the polymer can be delocalized above as well as below the interface. If $h > 0$ instead it can only be delocalized above: note that if $\lambda > 0$ and $\underline{v} \in \overset{\circ}{\mathcal{D}}$ then $h > 0$, as it has been pointed out in Section 6.3.2.

The proof is a consequence of the following lemma.

Lemma 8.2 *If $\underline{v} \in \overset{\circ}{\mathcal{D}}$ then*

$$\lim_{N \to \infty} \frac{1}{N} \mathcal{S} \left(\mathbf{P}^a_{N,\omega,\underline{v}} | \mathbf{P} \right) = 0, \qquad (8.5)$$

$\mathbb{P}(\,d\omega)-a.s..$

Proof. Let us consider $\underline{v}(t) := (\lambda(1 + t), h, \widetilde{\lambda}(1 + t), h(1 + t))$. By the hypothesis, $\mathrm{F}(\underline{v}(t)) = 0$ for t in a neighborhood of 0, in particular (by convexity)

$$0 = \partial_t \mathrm{F}(\underline{v}(t)) = \lim_{N \to \infty} \frac{1}{N} \partial_t \log Z^a_{N,\omega,\underline{v}(t)}\Big|_{t=0}, \qquad (8.6)$$

$\mathbb{P}(\,d\omega)-a.s..$ On the other hand, by direct computation

$$\mathcal{S} \left(\mathbf{P}^a_{N,\omega,\underline{v}} | \mathbf{P} \right) = \partial_t \log Z^a_{N,\omega,\underline{v}(t)}\big|_{t=0} - \log Z^a_{N,\omega,\underline{v}}, \qquad (8.7)$$

and the result follows. $\qquad \square$

Proof of Proposition 8.1. We first claim that if

(1) $\{E_N\}_N$ is a sequence of measurable subsets of Ω_S such that for $\varepsilon > 0$ we have the bound $\mathbf{P}(E_N) \leq \exp(-\varepsilon N)$ for every N;
(2) $\{\mathbf{P}_N\}_N$ is a sequence of measures such that $\mathbf{P}_N$ is equivalent to $\mathbf{P}$ for every N and such that $\mathcal{S}(\mathbf{P}_N | \mathbf{P}) = o(N)$;

then $\mathbf{P}_N(E_N) = o(1)$.

The claim is a direct consequence of (A.10), which yields for every $\alpha > 0$

$$\frac{1}{\alpha N} \log \mathbf{E} \exp (\alpha N 1_{E_N}) \geq \mathbf{P}_N (E_N) - \frac{1}{\alpha N} \mathcal{S} (\mathbf{P}_N | \mathbf{P}), \qquad (8.8)$$

from which

$$\mathbf{P}_N(E_N) \leq \frac{1}{\alpha N} \log \left(1 + \left(e^{\alpha N} - 1 \right) \mathbf{P}(E_N) \right) + \frac{1}{\alpha N} \mathcal{S} (\mathbf{P}_N | \mathbf{P}), \qquad (8.9)$$

and by choosing $\alpha = \varepsilon/2$ we obtain the claim.

We now specialize to $\mathbf{P}_N = \mathbf{P}_{N,\omega,\underline{v}}$ and $E_N = E_N(k) := \{S : \sum_{n=1}^N 1_{S_n=k} \geq \varrho N\}$ ($k \in \mathbb{Z}$ and $\varrho > 0$). Note that if we show that for every $\varrho > 0$ we have $\mathbf{P}_{N,\omega,\underline{v}}(E_N) = o(1)$, $\mathbb{P}(\,d\omega)$–a.s., we are done since it implies that

$$\lim_{N\to\infty} \frac{1}{N}\mathbf{E}_{N,\omega,\underline{v}}\left[\sum_{n=1}^{N}\mathbf{1}_{k_1\leq S_n\leq k_2}\right] = 0, \tag{8.10}$$

$\mathbb{P}(\,d\omega)$–a.s. for every $k_1, k_2 \in \mathbb{Z}$ and we already know (8.1) and (8.2).

By Lemma 8.2 the hypothesis on the relative entropy is satisfied for typical ω. In order to estimate the probability of $E_N(k)$ we observe that, for $k > 0$ and with the notation $\zeta = \inf\{n : S_n = k\}$ we have $\mathbf{P}(E_N(k)) = \sum_{j\geq k}\mathbf{P}(\zeta = j)\mathbf{P}\left(\mathcal{N}_{N-j}(\tau) + 1 \geq \varrho N\right) \leq \mathbf{P}(\mathcal{N}_N(\tau) \geq (\varrho N) - 1)$. Therefore it is sufficient to prove that $\mathbf{P}(\mathcal{N}_N(\tau) \geq \varrho N)$ vanishes exponentially fast for every $\varrho > 0$. We write for $\kappa > 0$

$$\mathbf{P}(\mathcal{N}_N(\tau) \geq \varrho N) = \mathbf{P}\left(\exp\left(-\kappa\tau_{\lceil\varrho N\rceil}\right) \geq \exp(-\kappa N)\right), \tag{8.11}$$

and we observe that

$$1 - \mathbf{E}\left[\exp(-\kappa\tau_1)\right] \stackrel{\kappa\searrow 0}{\sim} c_K\Gamma(1/2)\sqrt{\kappa} =: c\sqrt{\kappa}, \tag{8.12}$$

as it follows from Theorem A.2 and from $K(n) \stackrel{n\to\infty}{\sim} c_K/n^{3/2}$. Therefore, by the Markov inequality and the renewal property, we have

$$\frac{1}{N}\log\mathbf{P}(\mathcal{N}_N(\tau) \geq \varrho N) \leq \frac{\lceil\varrho N\rceil}{N}\log\mathbf{E}\left[\exp(-\kappa\tau_1)\right] + \kappa, \tag{8.13}$$

therefore there exist N_0 and κ_0 such that for $N \geq N_0$ and $\kappa \leq \kappa_0$ the right-hand side is bounded above by $-(c\varrho/2)\sqrt{\kappa}+\kappa$ which is negative for κ small. Since ϱ can be chosen arbitrarily small, the proof is complete.   Prop. 8.1
□

8.2   Beyond Large Deviations: $O(\log N)$ Results

We focus on the disordered pinning model. The changes that are needed to extend the results to the copolymer case are summed up in Section 8.4.

The main result of this section improves sensibly on (8.2) taking advantage of concentration properties of product measures. It has however the drawback of being a quenched averaged estimate.

Theorem 8.3   *Besides the standard assumptions on ω, let us assume that the concentration inequality (A.16) holds with $\vartheta(t) = c_1\exp(-c_2 t^p)$, for a*

$p \in (0, 2]$. *Then for every* $(\beta, h) \in \overset{\circ}{\mathcal{D}}$ *there exist two positive constants* c *and* q *such that*

$$\mathbb{E} \mathbf{P}^a_{N,\omega} \left(\mathcal{N}_N(\tau) \geq n \right) \leq \exp \left(-cn^{p/2} \right), \tag{8.14}$$

both for $a = \mathtt{c}$ *and* $a = \mathtt{f}$ *and for every* N *and* $n \geq q \log N$.

Particular cases in which Theorem 8.3 holds include ω_1 bounded or Gaussian (in both cases $p = 2$) and ω_1 bi-exponential, with $p = 1$, see Appendix A.3.

The proof of Theorem 8.3 is based on concentration inequalities for suitably restricted partition function, namely we will analyze $Z^a_{N,\omega}(\mathcal{N}_N(\tau) = n)$.

Set

$$\mathrm{F}^a_{N,\omega}(\beta, h; n) := \frac{1}{N} \log Z^a_{N,\omega} \left(\mathcal{N}_N(\tau) = n \right), \tag{8.15}$$

and $\mathrm{F}^a_{N,\omega}(\beta, h) := (1/N) \log Z^a_{N,\omega}$ We have the following sharpening of (4.33).

Lemma 8.4 *Under the same assumptions as in Theorem 8.3 we have that*

$$\mathbb{P} \left(\left| \mathrm{F}^a_{N,\omega}(\beta, h; n) - \mathbb{E} \left[\mathrm{F}^a_{N,\omega}(\beta, h; n) \right] \right| \geq u \right) \leq c_1 \exp \left(-c_2 \left(\frac{uN}{|\beta|\sqrt{n}} \right)^p \right), \tag{8.16}$$

for every $N \in \mathbb{N}$, $n \in \{0, 1, \ldots, N\}$ *and every* $u \geq 0$.

If $n = N$ we recover (4.33), but for small n the bound improves substantially.

Remark 8.5 *In the proof of Theorem 8.3 in reality it suffices the estimate (8.4) without the absolute value in the left-hand side, i.e. it is sufficient to control the deviations above the mean. Since deviations below the mean demand at times techniques that are different from the ones one employs for the deviations above the mean [Ledoux (2001)], this observation may have some relevance for disorders that are more general than the ones we consider here.*

Proof of Theorem 8.3. Let us choose $(\beta, h) \in \overset{\circ}{\mathcal{D}}$ and let us consider first the case $a = \mathtt{c}$. In this case we can exploit the fact that $\{N\mathbb{E}[\mathrm{F}^{\mathtt{c}}_{N,\omega}(\beta, h)]\}_N$

is super-additive (Proposition 4.2) so that $F(\lambda, h) = \sup_N \mathbb{E}[F^c_{N,\omega}(\beta, h)]$ by the standard property of super-additive sequences (Proposition A.12). This implies that

$$(\beta, h) \in \mathcal{D} \iff \mathbb{E}\left[F^c_{N,\omega}(\beta, h)\right] \leq 0 \text{ for every } N \in \mathbb{N}. \qquad (8.17)$$

Exploiting the fact that in reality we have chosen (β, h) in the interior of $\mathcal{D}$, we know that also $(\beta, h - \varepsilon) \subset \mathcal{D}$ for some $\varepsilon > 0$. Observe that for every ω we have the equivalence

$$F^c_{N,\omega}(\beta, h; n) \geq -\frac{1}{2}\varepsilon n/N \iff F^c_{N,\omega}(\beta, h - \varepsilon; n) \geq \frac{1}{2}\varepsilon n/N, \qquad (8.18)$$

and that $F^c_{N,\omega}(\beta, h - \varepsilon; n) \leq F^c_{N,\omega}(\beta, h - \varepsilon)$ which in turn has non-positive expectation, as we have seen in (8.17). Therefore

$$\mathbb{P}\left(F^c_{N,\omega}(\beta, h; n) \geq -\frac{1}{2}\varepsilon n/N\right) = \mathbb{P}\left(F^c_{N,\omega}(\beta, h - \varepsilon; n) \geq \frac{1}{2}\varepsilon n/N\right)$$

$$\leq \mathbb{P}\left(F^c_{N,\omega}(\beta, h - \varepsilon; n) - \mathbb{E}\left[F^c_{N,\omega}(\beta, h - \varepsilon; n)\right] \geq \frac{1}{2}\varepsilon n/N\right)$$

$$\leq c_1 \exp\left(-c_2 n^{p/2}\left(\frac{\varepsilon}{2|\beta|}\right)^p\right), \qquad (8.19)$$

where in the last step we have applied Lemma 8.4. Consider now the event $E_m := \{\omega : \text{ there exists } n \geq m \text{ such that } F^c_{N,\omega}(\beta, h; n) \geq -\varepsilon n/2N\}$. By (8.19) and by the union bound we have that there exists $C_1 > 0$ such that

$$\mathbb{P}(E_m) \leq \sum_{n \geq m} \mathbb{P}\left(F^c_{N,\omega}(\beta, h; n) \geq -\frac{1}{2}\varepsilon n/N\right) \leq C_1^{-1} \exp(-C_1 m^{p/2}),$$

$$(8.20)$$

for every m. Choose now $\omega \in E_m^C$. We have:

$$\mathbf{P}^c_{N,\omega}\left(\mathcal{N}_N(\tau) \geq m\right) = \frac{1}{Z^c_{N,\omega}} \sum_{n \geq m} Z^c_{N,\omega}\left(\mathcal{N}_N(\tau) = n\right)$$

$$\leq C_2 N^{1+\alpha_+} \exp\left(-\beta\omega_N + h\right) \sum_{n \geq m} \exp\left(-\varepsilon n/2\right) \qquad (8.21)$$

$$\leq C_3 N^{1+\alpha_+} \exp\left(-\beta\omega_N + h\right) \exp\left(-\varepsilon m/2\right),$$

where in the second line we have used the standard bound $Z^c_{N,\omega} \geq K(N)\exp(\beta\omega_N - h)$ and $\alpha_+ > \alpha$. By putting (8.20) and (8.21) together we obtain that there exists C_4 such that

$$\mathbf{EP}^c_{N,\omega}\left(\mathcal{N}_N(\tau) \geq m\right) \leq C_4^{-1} N^{1+\alpha_+} \exp(-C_4 m) + C_1^{-1} \exp(-C_1 m^{p/2}), \tag{8.22}$$

which yields the statement for the case $a = \mathtt{c}$.

If $a = \mathtt{f}$ instead we use (4.25), so that

$$\mathbb{E}\left[\mathrm{F}^{\mathtt{f}}_{N,\omega}(\beta, h)\right] \leq \frac{1}{N} \log(C_5 N), \tag{8.23}$$

for $(\beta, h) \in \mathcal{D}$ and the proof follows exactly like in the constrained case except that one has to replace $\mathrm{F}^c_{N,\omega}(\beta, h; n)$ with $\mathrm{F}^{\mathtt{f}}_{N,\omega}(\beta, h; n) - (\log(C_5 N))/N$. Moreover in (8.21) one uses $Z^{\mathtt{f}}_{N,\omega} \geq \sum_{n>N} K(n) \geq cN^{-\alpha_+}$ for any $\alpha_+ > \alpha$ and some $c > 0$. Since we are not tracking the precise values of the constants, the net result is the same.   Theorem 8.3 $\qquad\square$

Proof of Lemma 8.4. By (A.16) it is sufficient to show that

$$Q_{N,n}(\omega, \omega') := \left|\mathrm{F}^a_{N,\omega}(\beta, h; n) - \mathrm{F}^a_{N,\omega'}(\beta, h; n)\right| \leq \frac{|\beta|\sqrt{n}}{N}|\omega - \omega'|. \tag{8.24}$$

As in the proof of (4.33) we write $\omega_t := t\omega + (1-t)\omega'$ for $t \in [0,1]$ and we have

$$Q_{N,n}(\omega, \omega') = \left|\int_0^1 \frac{\beta}{N} \sum_{n=1}^N (\omega_n - \omega'_n)\, \mathbf{E}^a_{N,\omega_t}\left[\delta_n \middle| \mathcal{N}_N(\tau) = n\right] \mathrm{d}t\right|$$

$$\leq \frac{|\beta|}{N} \sqrt{\sup_t \sum_{k=1}^N \left(\mathbf{E}^a_{N,\omega_t}\left[\delta_k \middle| \mathcal{N}_N(\tau) = n\right]\right)^2 \sum_{n=1}^N (\omega_n - \omega'_n)^2}$$

$$\leq \frac{|\beta|}{N} \sqrt{\sup_t \sum_{k=1}^N \left(\mathbf{E}^a_{N,\omega_t}\left[\delta_k \middle| \mathcal{N}_N(\tau) = n\right]\right) \sum_{n=1}^N (\omega_n - \omega'_n)^2}$$

$$= |\beta|\frac{\sqrt{n}}{N} \sqrt{\sum_{n=1}^N (\omega_n - \omega'_n)^2}, \tag{8.25}$$

in which we have used the Cauchy–Schwarz inequality. Therefore (8.24) holds and the proof is complete.   Lemma 8.4 $\qquad\square$

8.3   The Strongly Delocalized Regime

Stronger delocalization results may be obtained *deeply* inside $\mathcal{D}$. By deeply inside $\mathcal{D}$ we mean in

$$\mathcal{D}_s = \{(\beta, h) : \mathbb{E}\left[\exp(\beta\omega_1 - h)\right] < 1/\Sigma_K\}. \tag{8.26}$$

By Jensen inequality $\mathcal{D}_s \subset \mathcal{D}$. The subscript s stands for *strongly delocalized* and the definition is naturally generalized to copolymers and copolymers with adsorption. Note that $(\beta, h) \in \mathcal{D}_s$ if and only if $h > \log \Sigma_K + \log M(\beta)$.

Theorem 8.6   *If $(\beta, h) \in \mathcal{D}_s$ then there exists $c > 0$ such that for every N*

(1) we have

$$\mathbb{E}\mathbf{P}^a\left(\mathcal{N}_N(\tau) \geq n\right) \leq c\exp\left(-n/c\right), \tag{8.27}$$

for every n and both for $a = \mathsf{c}$ and $a = \mathsf{f}$;
(2) we have

$$\mathbb{E}\mathbf{P}^{\mathsf{f}}_{N,\omega}\left(\max(\tau \cap [0, N]) \geq n\right) \leq c\overline{K}(n), \tag{8.28}$$

for every n and

$$\mathbb{E}\mathbf{P}^{\mathsf{c}}_{N,\omega}\left(\max(\tau \cap [0, N/2]) \geq n_1, \min(\tau \cap [N/2, N]) \leq n_2\right) \\ \leq c\overline{K}(n_1)\overline{K}(N - n_2), \tag{8.29}$$

for every n_1 and n_2.

Proof. By the Fubini–Tonelli Theorem we have

$$\mathbb{E}\left[Z^{\mathsf{c}}_{N,\omega}\left(\mathcal{N}_N(\tau) = n\right)\right]$$
$$= \mathbf{E}\left[\exp\left(\left(\hat{\beta} - \log\Sigma_K\right)\mathcal{N}_N(\tau)\right); \mathcal{N}_N(\tau) = n, N \in \tau\right]$$
$$= \sum_{\substack{\ell \in \mathbb{N}^n: \\ \sum_{j=1}^n \ell_j = N}} \prod_{j=1}^n \left(\frac{K(\ell_j)}{\Sigma_K}\exp\left(\hat{\beta}\right)\right) \tag{8.30}$$
$$= \tilde{\mathbf{E}}\left[\exp\left(\hat{\beta}\mathcal{N}_N(\tau)\right); \mathcal{N}_N(\tau) = n, N \in \tau\right],$$

where $\hat{\beta} = \log \mathrm{M}(\beta) - h + \log \Sigma_K$ and $\widetilde{\mathbf{P}}$ is the law of the renewal with inter-arrival distribution $\widetilde{K}(\cdot) := K(\cdot)/\Sigma_K$. Note that in these steps we have applied the idea set forth in Remark 1.19. Moreover we have

$$
\begin{aligned}
\mathbb{E}\left[Z^{\mathrm{c}}_{N,\omega}\left(\mathcal{N}_N(\tau) = n\right)\right] &= \exp\left(\hat{\beta}n\right)\widetilde{\mathbf{P}}\left(\mathcal{N}_N(\tau) = n, \ N \in \tau\right) \\
&= \exp\left(\hat{\beta}n\right)\widetilde{K}^{n*}(N) \leq \exp\left(\hat{\beta}n\right)n^c\widetilde{K}(N),
\end{aligned}
\tag{8.31}
$$

where in the last step we have applied Lemma A.5 and $c > 0$ is the constant appearing in that statement. Therefore if $(\beta, h) \in \mathcal{D}_{\mathbf{s}}$, that is $\hat{\beta} < 0$, we have

$$
\begin{aligned}
\mathbb{E}\mathbf{P}^{\mathrm{c}}\left(\mathcal{N}_N(\tau) \geq n\right) &= \sum_{m \geq n} \mathbb{E}\left[\frac{Z^{\mathrm{c}}_{N,\omega}\left(\mathcal{N}_N(\tau) = m\right)}{Z^{\mathrm{c}}_{N,\omega}}\right] \\
&\leq \frac{1}{K(N)} \sum_{m \geq n} \mathbb{E}\left[\exp(-\beta\omega_N + h)Z^{\mathrm{c}}_{N,\omega}\left(\mathcal{N}_N(\tau) = m\right)\right] \\
&\leq \frac{\exp\left(-\log \mathrm{M}(\beta) + h\right)}{K(N)} \sum_{m \geq n} \mathbb{E}\left[Z^{\mathrm{c}}_{N,\omega}\left(\mathcal{N}_N(\tau) = m\right)\right] \\
&\leq \frac{\exp\left(-\log \mathrm{M}(\beta) + h\right)}{\Sigma_K} \sum_{m \geq n} m^c \exp\left(\hat{\beta}m\right),
\end{aligned}
\tag{8.32}
$$

and (8.27) is proven for $a = \mathrm{c}$.

Considering still the constrained case, call E_{n_1, n_2} the event in the left-hand side of (8.29). The same steps as in (8.30) apply and we obtain

$$
\mathbb{E}\left[Z^{\mathrm{c}}_{N,\omega}\left(E_{n_1,n_2}\right)\right] = \widetilde{\mathbb{E}}\left[\exp\left(\hat{\beta}\mathcal{N}_N(\tau)\right); \ E_{n_1,n_2} \cap \{N \in \tau\}\right],
\tag{8.33}
$$

and the expression in the right-hand side is the partition function, restricted to E_{n_1, n_2}, of a homogeneous pinning model with recurrent inter-arrivals, *i.e.* in the delocalized (non-critical) regime as soon as $\hat{\beta}$ is negative. At this point we choose $\hat{\beta} < 0$ and observe that

$$
\begin{aligned}
\mathbb{E}\mathbf{P}^{\mathrm{c}}_{N,\omega}\left(E_{n_1,n_2}\right) &\leq \frac{\exp\left(-\log \mathrm{M}(\beta) + h\right)}{K(N)}\mathbb{E}\left[Z^{\mathrm{c}}_{N,\omega}\left(E_{n_1,n_2}\right)\right] \\
&\leq c\widetilde{\mathbf{P}}^{\mathrm{c}}_{N,\hat{\beta}}\left(E_{n_1,n_2}\right),
\end{aligned}
\tag{8.34}
$$

where $c > 0$ and we have used (8.33) and Theorem 2.2(2) in order to substitute $K(N)$ with the partition function of a constrained endpoint model with pinning $\hat{\beta}$ and inter-arrival distribution $\widetilde{K}(\cdot)$. The estimate we need is therefore reduced to an estimate on a homogeneous model: this has been already done in Chapter 2, (2.44) (see also Remark 2.6) and (8.29) is therefore proven.

The proof in the free endpoint case is very similar. Still for $(\beta, h) \in \mathcal{D}_s$, i.e. $\hat{\beta} < 0$, we have

$$\mathbb{E}\left[Z_{N,\omega}^f\left(\mathcal{N}_N(\tau) = n\right)\right] \leq n^c \exp\left(\hat{\beta}n\right) \sum_{m=0}^{N-n} \widetilde{K}(N-m)\left[\overline{K}(m) + K(\infty)\right]$$

$$\leq Cn^c \exp\left(\hat{\beta}n\right)\left[\overline{K}(N) + K(\infty)\right],$$

$$(8.35)$$

as a direct consequence of the standard formula linking free case and constrained case and of (8.31). In the last step, if $K(\infty) = 0$, we have used the fact that $\sum_{m=0}^{N-1} \widetilde{K}(N-m)\overline{K}(m) \leq C'\overline{K}(N)$, for some positive constant C'. This follows immediately for $\alpha > 0$ by splitting the sum according to whether m is larger or smaller than $N/2$ and by making rough estimates. If $\alpha = 0$ one makes the same splitting but then one has to estimate the terms a bit more carefully by using L.4 of Appendix A.4. At this point we proceed like in (8.32) to obtain (8.27) in the case $a = f$. The bound (8.28) is obtained instead by proceeding like in (8.34) and by applying (2.45). □

We give now a scaling limit result in the strongly delocalized regime under the condition that $K(\cdot)$ decays sufficiently rapidly. One is certainly tempted to conjecture the validity of such a result in the whole class of inter-arrival distributions that we consider and everywhere in $\mathcal{D}$. We will come back to this issue in the next section (with no definite answer).

Corollary 8.7 *For every N consider the law $\mu_{N,\omega}^a$ of $\tau_{(N)} := (\tau/N) \cap [0, 1]$ (τ is distributed according to $\mathbf{P}_{N,\omega}^a$), seen as a measure on the closed subsets of $[0, 1]$ endowed with the Hausdorff metric (see Appendix A.5.4). If $\sum_{n \in \mathbb{N}} nK(n) < \infty$ and if $(\beta, h) \in \mathcal{D}_s$ then $\mathbb{P}(\mathrm{d}\omega)$–a.s. we have that the $\mu_{N,\omega}^f \stackrel{N \to \infty}{\Longrightarrow} \delta_{\{0\}}$ and $\mu_{N,\omega}^c \stackrel{N \to \infty}{\Longrightarrow} \delta_{\{0,1\}}$.*

Proof. Since $\sum_{n \in \mathbb{N}} nK(n) = \sum_{n=0}^{\infty} \overline{K}(n) < \infty$ (recall that $K(\infty)$ may or may not be equal to zero) we can find $g : \mathbb{N} \to \mathbb{N}$ such that $\lim_{n \to \infty} g(n) = +\infty$, $g(n) = o(n)$ and $\sum_n \overline{K}(g(n)) < \infty$. In particular if $\alpha > 1$ one may choose $g(n)$ equal to the integer part of n^{1/α_-}, $1 < \alpha_- < \alpha$. Therefore in the free endpoint case, by (8.28) we have

$$\sum_N \mathbb{E}\mathbf{P}^{\mathrm{f}}_{N,\omega}\left(\max \tau_{(N)} \geq g(N)/N\right) < \infty, \tag{8.36}$$

which, by the Borel–Cantelli argument (Appendix A.1), says that

$$\lim_{N \to \infty} \mathbf{P}^{\mathrm{f}}_{N,\omega}\left(\max \tau_{(N)} \geq g(N)/N\right) = 0, \tag{8.37}$$

$\mathbb{P}(\,\mathrm{d}\omega)$–a.s.. Since $g(N)/N = o(1)$ we are done. □

8.4   Bibliographic Complements

Complements to Section 8.1

The results generalize to our context the treatment given in [Biskup and den Hollander (1999)] for copolymers.

Complements to Section 8.2

The results of this section are taken from [Giacomin and Toninelli (2005)], where they are worked out in the copolymer setting with hypothesis of Gaussian concentration ($p = 2$ in Theorem 8.3). The argument is absolutely parallel to the case of pinning presented here: one has simply to replace $\mathcal{N}_N(\tau) = \sum_{n=1}^N \delta_n$ with $\sum_{n=1}^N \Delta_n$ in the proof. The statement holds naturally with $\sum_{n=1}^N \Delta_n$ instead of with $\mathcal{N}_N(\tau)$ and may be extended to $\mathcal{N}_N(\tau)$ without much trouble.

By coupling the techniques developed in this section and the idea at the base of the smoothing inequality of Section 5.4, F. L. Toninelli has obtained an estimate on the contact fraction at and near criticality, which (for example, and roughly) says that there at most $N^{2/3} \log N$ pinned sites at criticality, regardless of the value of α. For a precise statement we refer to [Toninelli (2006)]. Here we remark that the procedure is based on replacing the crucial ingredient (8.17), which holds inside the delocalized phase, with the quadratic upper bound that one obtains with the techniques of Section 5.4 (hence these critical estimates are probably not optimal, at least not in all regimes, see Sec 5.5).

Complements to Section 8.3

This section generalizes [Giacomin and Toninelli (2005)].

A general consideration on this chapter is that the results presented are sensibly weaker than those in Chapter 2 and Chapter 3 for homogeneous and weakly inhomogeneous models. It is then natural to ask: what should we expect to observe in the delocalized regime of disordered models? This question is particularly relevant also because weakly inhomogeneous models have been studied as caricatures of disordered models.

We make this question more precise by considering the following two aspects:

(1) Should one expect that for $(\beta, h) \in \overset{\circ}{\mathcal{D}}$ the visits to the interface (the contacts) be all at a finite distance from the boundary? More precisely we are asking whether or not, say in the free endpoint case, $\mathbf{P}(\,d\omega)$-a.s. the sequence, indexed by N, of the distance from the origin of the rightmost contact with the interface is tight. In alternative one could ask whether a weaker statement, like the one in Corollary 8.7, holds. Note that for such a result it suffices to prove that the contacts are at a distance $o(N)$ from the boundary.

(2) Should one expect the decay of $Z_{N,\omega}^{a}$, for $(\beta, h) \in \overset{\circ}{\mathcal{D}}$ and for typical ω, be like $K(N)$ if $a = \mathsf{c}$? In the free case $K(N)$ is replaced by $\overline{K}(N)$ if $K(\infty) = 0$, and by a positive constant otherwise (*cf.* Theorem 2.2(2)).

We do not really have satisfactory answers to these questions: but we have the following observations:

- In the strongly delocalized regime, from Theorem 8.6 one can extract the weak convergence statements of Corollary 8.7 for general return time distribution, but only along subsequences.
- On the other hand for every $(\beta, h) \in \mathcal{D}$ there exists $\varepsilon > 0$ such that one can find arbitrarily large values of N such that

$$Z_{N,\omega}^{\mathsf{c}} \geq K(N)N^{\varepsilon}. \tag{8.38}$$

In particular instances this result implies a (very weak) result on the trajectories indicating that one cannot extend to the disordered set-up in the most naive way the results obtained in the homogeneous and weakly inhomogeneous set-up.

We will not detail further the first point since it suffices to have a look at the proof of Corollary 8.7 to realize that one can select a subsequence having the desired property. We turn instead to the second issue and we will discuss it in an informal way: we develop the idea set forth in [Giacomin and Toninelli (2005), Section 4].

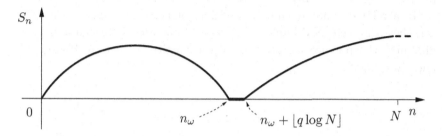

Fig. 8.1 We compute a lower bound on the free energy by selecting only the trajectories coming back to the interface only when the sequence of charges is atypical and it guarantees a large gain.

We choose for this discussion $\omega_1 \sim \mathcal{N}(0,1)$, select $(\beta, h) \in \mathcal{D}$ and $\delta > 0$ such that $\mathrm{F}(\beta, h - \delta) > 0$. With reference to Figure 8.1, we observe that if we have a polymer of length N for any $q < 1/\Sigma(\delta/\beta)$ ($\Sigma(\cdot)$ is the Cramér functional, *cf.* Appendix A.1), with probability close to 1 as $N \to \infty$, we find n_ω such that $Z^{\mathrm{c}}_{n_\omega, n_\omega + \lfloor q \log N \rfloor} \geq \exp\left(\mathrm{F}(\beta, h - \delta) \lfloor q \log N \rfloor / 2\right)$. This can be proven by exploiting the strategy developed in Section 5.4: the idea behind is simply that one can find a stretch of charges of length $\lfloor q \log N \rfloor$ with empirical mean smaller than h and therefore a polymer pinned exactly at the endpoints of this stretch is approximately a polymer with charges $\beta \omega_n - h$ (in order to prove this one has to couple the analysis in [Dembo and Zeitouni (1998), Section 3.2] with the entropy argument in Section 5.4). We can now write a lower bound on $Z^{\mathrm{c}}_{n_\omega + \lfloor q \log N \rfloor, \omega}$ by selecting only the trajectories that touch the axis for the first time in n_ω: this yields the lower bounds

$$Z^{\mathrm{c}}_{n_\omega + \lfloor q \log N \rfloor, \omega} \geq K(n_\omega) \exp\left(\mathrm{F}(\beta, h - \delta) \lfloor q \log N \rfloor / 2\right), \qquad (8.39)$$

with probability close to one. One then obtains (8.38) by replacing $n_\omega + \lfloor q \log N \rfloor$ with N (and using the asymptotic monotonicity of $K(\cdot)$).

While this argument yields a very weak result, it shows that things cannot be as *easy* as in the homogeneous case. We refer to [Giacomin and Toninelli (2005), Section 4] for a (more detailed) argument close in spirit to what we have developed above. That argument works only for copolymers, while the argument given here is absolutely general (the pinning set-up has been chosen for notational convenience), and it says (roughly) that in the delocalized (non-critical) regime of the copolymer model, but for $h < h^{(1)}(\lambda)$, the $\mathbf{P}^{\mathrm{f}}_{N,\omega}$ probability that there is a contact at distance $(\log N)^c$, $c \in (0, 1)$, (or farther) from the origin does not tend to zero $\mathbb{P}(\mathrm{d}\omega)$–a.s..

In a related context, namely modeling of RNA secondary structures, a replica argument and numerical computations are performed aiming at obtaining the polynomial behavior of the partition function [Bundschuh and Hwa (2002)].

Chapter 9

Numerical Algorithms and Computations

Random directed polymers still present a number of open questions and this gives impulse to attack them from a numerical viewpoint. The one-dimensional character of the problems allows the numerical treatment of *very large* systems.

In this chapter we give some algorithms for computing partition functions, along with some applications. We insist that we are really computing partition functions, a procedure that is virtually impossible in most of the cases of interest in statistical mechanics and one in fact resorts to the dynamical Monte Carlo approach precisely to get around the difficulty of computing the partition function (see [Madras (2002)] for an introduction to these techniques). We instead use random number generators only to generate the disorder variables and, once the disorder variables are given, the partition function is computed exactly (up to the precision of the machine).

We, however, have to stress that the chapter focuses only on computations with a constant look at rigorous results. The computation is mainly seen as a tool for getting ideas on what to prove and how to prove it and therefore it rarely pushes much beyond the boundary of what has been rigorously understood. In Section 9.4 we will refer to works pushing further, trying for example to estimate the critical exponents of the various models, but this will not be considered in detail.

One of the main aims of this chapter is to explain that one can assert localization by numerical computation in an *almost rigorous* way, while asserting delocalization is essentially out of control.

9.1   Computing the Partition Function

This section is split in three parts dealing with:

(1) models based on random walks (so $\alpha = 1/2$ and $L(\cdot)$ is trivial);
(2) models with general return times;
(3) a fast algorithm for the model with return times distribution $K(n) = c/n^{1+\alpha}$, $c > 0$.

9.1.1   *Random walk models*

Let us consider the case of models based on the simple random walk (that is the $(1,0)$-walk: the generalization to (p,q)-walks is immediate). The aim of this section is to explain the *transfer-matrix* method for computing $Z_{N,\omega}^{a}$ and we are going to see that $Z_{N,\omega}^{a}$ can be computed in $O(N^2)$ steps. The notation $O(\cdot)$ is the standard one used in computer science and, in a sense, it coincides with the notation used in mathematical analysis (and throughout this book).

And precisely about the speed of computation, observe that computing directly $Z_{N,\omega}^{f}$ would require summing up 2^N terms, the number of random walk trajectories, making the problem computationally out of reach (and of course the situation is absolutely analogous for $Z_{N,\omega}^{c}$). However, here we can take profit of the *additivity* of the Hamiltonians we consider: loosely speaking, if we join together two (finite) random walk segments, the energy of the resulting path is the sum of the energies of the building segments. This fact can be exploited to derive a recurrence relation for the sequence $\{\mathcal{Z}_M(y)\}_{M=0,1,\dots}$, with $y \in \mathbb{Z}$ and $\mathcal{Z}_M(y) := Z_{2M,\omega}^{c}(S_{2M} = 2y)$. Conditioning on S_{2M} and using the Markov property we directly find for the pinning model

$$\mathcal{Z}_{M+1}(y) = \begin{cases} \frac{1}{4}\left[\mathcal{Z}_M(y+1) + 2\mathcal{Z}_M(y) + \mathcal{Z}_M(y-1)\right] & \text{if } y \neq 0 \\ \frac{1}{4}\left[\mathcal{Z}_M(1) + 2\mathcal{Z}_M(0) + \mathcal{Z}_M(-1)\right] \exp\left(\beta\omega_{2M+2} - h\right) & \text{if } y = 0. \end{cases}$$
$$(9.1)$$

In the copolymer case we have instead

$$\mathcal{Z}_{M+1}(y) = \begin{cases} \frac{1}{4}\left[\mathcal{Z}_M(y+1) + 2\mathcal{Z}_M(y) + \mathcal{Z}_M(y-1)\right] & \text{if } y > 0 \\ \frac{1}{4}\left[\mathcal{Z}_M(1) + \mathcal{Z}_M(0)\right] + \frac{1}{4}r_M\left[\mathcal{Z}_M(0) + \mathcal{Z}_M(-1)\right] & \text{if } y = 0 \\ \frac{1}{4}r_M\left[\mathcal{Z}_M(y+1) + 2\mathcal{Z}_M(y) + \mathcal{Z}_M(y-1)\right] & \text{if } y < 0, \end{cases}$$
$$(9.2)$$

where we have set $r_M := \exp\left(-2\lambda\left(\omega_{2M+1} + \omega_{2M+2} + 2h\right)\right)$ and we have considered the model in which $\text{sign}(S_{2M}) = \text{sign}(S_{2M-1})$ if $S_{2M} = 0$.

Of course $\mathcal{Z}_0(y) = \mathbf{1}_{y=0}$ and $\mathcal{Z}_M(y) = 0$ if $|y| > M$, so from the recursions (9.1) and (9.2) it follows that $\{\mathcal{Z}_{M+1}(y)\}_{y\in\{-M-1,\dots,M+1\}}$ can be obtained from $\{\mathcal{Z}_M(y)\}_{y\in\{-M,\dots,M\}}$ with $O(M)$ computations. This means that we can compute $Z_{N,\omega}^a$, both for $a = \mathsf{c}$ and $a = \mathsf{f}$, in $O(N^2)$ steps. Note that $\mathcal{Z}_M(x)/\sum_y \mathcal{Z}_M(y)$ is equal to $\mathbf{P}_{N,\omega}^{\mathsf{f}}(S_N = x)$, $N = 2M$, that is the discrete density of the distance from the interface of the last monomer of the chain and several other probability distributions can be easily extracted from $\{\mathcal{Z}_M(y)\}_{M,y}$. The algorithm just described can be implemented in a standard way (implementations of it, written in C, are available on the web page of the author).

We point out that sometimes one is satisfied with *lower bounds* on $Z_{N,\omega}^a$, for instance in the statistical test for localization described in Section 9.2. In this case the algorithm can be further speeded up by restricting the computation to a suitable set of random walk trajectories: instead of summing, at the Mth step of the iteration, over $y \in \{-M-1,\dots,M+1\}$ one can sum over a subset. Our understanding of the polymer behavior suggests that if the system size is N the endpoint of the polymer is typically at distance $O(\sqrt{N})$ in all regimes (and $\max_{n\le N}|S_n| = O(N^{(1/2)+\varepsilon})$, any $\varepsilon > 0$), hence a natural choice to get a lower bound on the partition function is to take into account only the contribution coming from the random walk paths $\{S_n\}_{n\in\mathbb{N}}$ for which

$$-An^{1/2} \le S_n \le Bn^{1/2} \quad \text{for } n \ge N_0\,, \tag{9.3}$$

where A, B, N_0 are positive constants. Of course this is easily implemented in the algorithm described above: it suffices to restrict the sums to $y \in [-A\sqrt{M}, B\sqrt{M}] \cap \mathbb{Z}$, while setting $\mathcal{Z}_{M+1}(y) = 0$ for the other values of y. In this way the number of computations needed to obtain Z_N is reduced to $O(N^{3/2})$. Specific values of A, B and N_0 will be given in Section 9.3.

9.1.2   *General renewal models*

For general renewal models the recursion is even simpler to write. We consider directly the case of copolymers with adsorption of (1.66), but of course in the general return time framework: $Z_{0,\omega}^c = 1$ and

$$Z_{N+1,\omega}^c = e^{\beta\omega_{N+1}-\widetilde{h}} \sum_{n=0}^{N} K(N+1-n) \left(\frac{1 + e^{\lambda \sum_{j=n+1}^{N+1}(\hat{\omega}_j+h)}}{2} \right) Z_{n,\omega}^c. \quad (9.4)$$

And one can then compute $Z_{N,\omega}^f$, if $\{Z_{n,\omega}^c\}_{n=1,\ldots,N}$ is known. This iteration allows to compute $Z_{N,\omega}^a$ in $O(N^2)$ computational steps, but in practice it turns out to be less manageable than the one of the previous subsection, starting with the fact that it is not clear how to impose an effective cut-off like the one in (9.3). In a sense it is always possible to put a cut-off on the length of the excursion and this turns out to be extremely effective in the localized regime, as long as the cut-off is (much) larger than the scale of the typical excursions. This approximation is however out of control close of criticality and of course it has little meaning in the delocalized regime.

9.1.3   *The Fixman–Freire Algorithm*

We present now an algorithm that allows to compute, in an approximate fashion, the partition function in *not much more than* $O(N)$ computations. We give it for conciseness in the case of pinning. We make the choice $K(n) = C_K/n^{1+\alpha}$, C_K is a positive constant, for every $n \in \mathbb{N}$ (the algorithm can be extended in principle to more general functions, but the estimates we give depend on the specific choice of $K(\cdot)$ and the exact power law dependence appears to be particularly suited for the procedure that we are going to outline). For the pinning model (9.4) reads

$$Z_{N+1,\omega}^c = e^{\beta\omega_{N+1}-h} \sum_{n=0}^{N} K(N+1-n) Z_{n,\omega}^c. \quad (9.5)$$

We state the following result:

Proposition 9.1   *Fix $\alpha \geq 0$ and $C_K > 0$ so that $K(\cdot)$ is determined. For every N and every $\varepsilon > 0$ there exist $I \in \mathbb{N}$ and positive numbers $a_1, \ldots, a_I, b_1, \ldots, b_I$ such that*

$$\left| K(n) - \sum_{i=1}^{I} a_i \exp(-b_i n) \right| \leq \varepsilon K(n), \quad (9.6)$$

for every $n \in \{1, \ldots, N\}$. The choice of I, given N, may be made such that the application $N \mapsto I$ is $O(\log N)$.

This result follows directly from [Beylkin and Monzón (2005), Theorem A.1], where an explicit procedure for determining the approximation coefficients is given. But beyond the quantitative aspects of Proposition 9.1, let us replace $K(n)$ with $\widehat{K}_N(n) := \sum_{i=1}^{I} a_i \exp(-b_i n)$, and denote by $\widehat{Z}_{N,\omega}$ the partition function of the model with this new (N–dependent) return distribution, and let us go back to (9.5) to observe that we can write

$$\widehat{Z}_{N+1,\omega} = \zeta_{N+1} \sum_{i=1}^{I} \sum_{n=0}^{N} a_i \exp(-b_i(N+1-n)) \widehat{Z}_{n,\omega}, \qquad (9.7)$$

with $\zeta_n := \exp(\beta \omega_n - h)$. Therefore by setting

$$\mathcal{Q}_i(N) := \sum_{n=0}^{N-1} \exp(-b_i(N-n)) \widehat{Z}_{n,\omega}, \qquad (9.8)$$

we have

$$\mathcal{Q}_i(N+1) = \exp(-b_i) \left[\mathcal{Q}_i(N) + \zeta_N \sum_{j=1}^{I} a_j \mathcal{Q}_j(N) \right], \qquad (9.9)$$

to which we add the *boundary condition* $\mathcal{Q}_i(1) := a_i \exp(-b_i)$ obtaining thus a recursion that allows to compute $\widehat{Z}_{N,\omega}$ in $O(I \times N)$ steps (that is $O(N \log N)$ if we exploit the approximation proposed in Proposition 9.1). The iteration (9.9) goes normally under the name of *Fixman–Freire* scheme (see Section 9.4).

The problem of giving quantitative bounds on $Z_{N,\omega}^{c} - \widehat{Z}_{N,\omega}$, starting for example from a statement like Proposition 9.1, will not be considered here. All the same we remark that if $\widehat{K}_N(n) \leq K(n)$ for $n = 1, \ldots, N$ then $Z_{N,\omega}^{c} \geq \widehat{Z}_{N,\omega}$ and such an observation makes clear the interest of the Fixman–Freire scheme in view of establishing localization.

The recursion scheme for copolymer models (with or without adsorption) is conceptually analogous, even if it is technically slightly more involved (see [Garel and Monthus (2005c)]).

9.2   A Statistical Test for Localization

As we have seen $\{\mathbb{E} \log Z_{N,\omega}^{c}\}_N$ is a super-additive sequence, *cf.* Theorem 4.2, and, by Proposition A.12, one has the following characterization of the localized phase:

$$\mathrm{F} > 0 \iff \text{there exists } N \text{ such that } \mathbb{E}\log Z_{N,\omega}^{c} > 0. \qquad (9.10)$$

What interests us in this section is that if we can find one value of N for which $\mathbb{E}\log Z_{N,\omega}^{c} > 0$ then the system is localized.

Let us give a simple application of such an idea: for the copolymer model, based on the simple random walk, where one easily computes

$$\log Z_{2,\omega}^{c} = \log\left(\frac{1}{2} + \frac{1}{2}\exp\left(-2\lambda\left(\omega_1 + \omega_2 + 2h\right)\right)\right), \qquad (9.11)$$

and therefore, considering for example $\mathbb{P}(\omega_1 = +1) = \mathbb{P}(\omega_1 = -1) = 1/2$ (below, charges of this type are simply called *binary charges*), we have that

$$\sum_{\omega_1 = \pm 1, \omega_2 = \pm 1} \log\left(\frac{1}{2} + \frac{1}{2}\exp\left(-2\lambda\left(\omega_1 + \omega_2 + 2h\right)\right)\right) > 0, \qquad (9.12)$$

implies that $(\lambda, h) \in \mathcal{L}$. This of course does not yield exceptional information, but it says for example that $h_c(\lambda) > (1 - (1.171/\lambda))$ for λ sufficiently large (to be compared, *cf.* Theorem 6.1, with $(1 - h^{(2/3)}(\lambda))/\lambda \overset{\lambda \to \infty}{\to}$ $(3/4)\log 2 \approx 0.5198\ldots)$ and this might look promising (after all, the computation has been made only for $N = 2$). Notice however that, even in the simplest case of binary charges (to which we will stick in this section) $\mathbb{E}\log Z_{N,\omega}^{c}$ is the sum over 2^N terms. Of course in principle, for example in order to improve on $h_c(\lambda) \geq h^{(2/3)}(\lambda)$ it may be sufficient to compute $\mathbb{E}\log Z_{N,\omega}^{c}$ for $N = 10$ or thereabout (one cannot go much beyond anyway). We will come back to this very interesting issue, but we anticipate that one needs rather N of the order of 10^3 (which is a remarkably small number, but 2^{1000} is not).

What we propose is to *decide* whether $\mathbb{E}\log Z_{N,\omega}^{c}$ is positive by Monte Carlo sampling $\log Z_{N,\omega}^{c}$. Of course, by the law of large numbers, the empirical average of n independent copies of $\log Z_{N,\omega}^{c}$, that we denote by $\widehat{u}_n(N)$, is an asymptotic estimator of $\mathbb{E}\log Z_{N,\omega}^{c}$ but then one needs to decide how large n should be to decide that $\widehat{u}_n(N) > 0$ really implies $\mathbb{E}\log Z_{N,\omega}^{c} > 0$. In order to tackle this point we resort to concentration inequalities.

In this case it is of course important to keep track of the constants. We are then going to use (A.17), keeping in mind (4.32), and we obtain that for $t \geq 0$

$$\mathbb{P}\left(\frac{1}{n}\sum_{j=1}^{n}\left(\log \widetilde{Z}^{c}_{N,\theta^{jN}\omega} - \mathbb{E}\log \widetilde{Z}^{c}_{N,\omega}\right) \geq t\right) \leq \exp\left(-\frac{nt^2}{4\lambda^2 N}\right), \quad (9.13)$$

where of course the translated charges $\theta^{jN}\omega$ are just a trick to sum independent copies of the system. Note that we are using $\widetilde{Z}^{c}_{N,\omega}$ and not $Z^{c}_{N,\omega}$, see (1.60): according to the application this may or may not be a good choice, but note that the choice we have made yields the constant 4 in the right-hand side of (9.13), while the alternative choice yields 16. Our procedure is now the following: let us (numerically) sample n copies of $\log \widetilde{Z}^{c}_{N,\omega}$, for a fixed N, and compute $\widehat{u}_n(N)$. Le us also make the hypothesis (that conforming to the statistical literature we call H0, or *null hypothesis*)

$$\text{H0}: \quad \mathbb{E}\log \widetilde{Z}^{c}_{N,\omega} \leq 0. \quad (9.14)$$

Then (9.13) puts a bound on the probability that H0 be really correct, namely:

Proposition 9.2 *If $\widehat{u}_n(N) > 0$ we refuse H0 with a level of error which is at most* $\exp\left(-n(\widehat{u}_n(N))^2/(4\lambda^2 N)\right)$.

Of course, refusing H0 means $(\lambda, h) \in \mathcal{L}$.

We want also to stress that, since (9.13) holds also for deviations below the mean and not only above, Proposition 9.2 is *reversible*, in the sense that if $\widehat{u}_n(N) < 0$ we can refuse the (new) null hypothesis $\mathbb{E}\log Z^{c}_{N,\omega} > 0$ with the very same level of error and therefore conclude that $\mathbb{E}\log Z^{c}_{N,\omega} \leq 0$, which however does not imply $(\lambda, h) \in \mathcal{D}$. However the information that it gives should not be neglected: it is telling us that, for the given λ and h, it is not possible to assert localization at the size N, even if the (infinite volume) system is effectively localized. This will be exploited in the next section in order to investigate the possibility of improving the lower bound on the critical curve given in Theorem 6.1 by a computer assisted proof. It has also another interesting application because it gives a (statistical) lower bound on the correlation length of the system, as we are going to discuss next.

9.3   Applications: A Few Examples

The problems left open or even suggested by some of the theorems we have proven are virtually uncountable. Here we just quickly attack the problem of the *gap* in (6.8) (Theorem 6.1) and the question of going beyond the free energy lower bound in Section 5.2.

We have applied the statistical test on localization, that is Proposition 9.2 to the copolymer model based on simple random walk and with binary charges: we have chosen $\lambda = 0.8$, $h = 0.4714439 \approx h^{(0.7)}(0.8)$ (for comparison $h^{(2/3)}(0.8) \approx 0.455115$ and $h^{(1)}(0.8) \approx 0.5917538$ and of course we are in the region which is not covered by the bound (6.8)). We have used the algorithm built on the recursion (9.2). For example with n (the sample size) equal to 29000 we have obtained

$$\widehat{u}_{29000}(20000) \; = \; 5.088486\ldots \tag{9.15}$$

which, by Proposition 9.2, implies that for the chosen values of λ and h the system is localized with a level of error of less than $0.5 \cdot 10^{-6}$. The size of the system ($N = 20000$) and the number of samples is rather small and (9.15) has been obtained in less than one hour of computation (on a single 2 MHz G5 CPU). As a matter of fact one can easily go much beyond and find that $\widehat{u}_{100000}$ is $4.3633\ldots$ and the level of error then reduces to less than 10^{-16}.

As already set forth in the previous section, Proposition 9.2 can be exploited to get another interesting piece of information. Still for the copolymer with binary charges, the same of the previous paragraph, we choose $\lambda = 1$ and $h = 0.5305231 \approx h^{(0.66667)}(1)$ that is we place ourselves just above the lower bound $h^{(2/3)}(1)$ on the critical point for $\lambda = 1$. We ask ourselves how large should N be to observe $\mathbb{E} \log \widetilde{Z}^{\mathrm{c}}_{N,\omega} > 0$. Note that we do not have a proof that $N \mapsto \mathbb{E} \log \widetilde{Z}^{\mathrm{c}}_{N,\omega}$ is increasing, we have actually numerical evidence that it is not, but we have also numerical evidence that it is not increasing only for *small* values of N. In any case, we have obtained $\widehat{u}_{10^7}(3050) = -0.1373\ldots$, which, by Proposition 9.2, yields that for $N = 3050$ we have $\mathbb{E} \log \widetilde{Z}^{\mathrm{c}}_{N,\omega} < 0$ with a level of error of less that $1.5 \cdot 10^{-7}$ (and we have obtained results in accord with monotonicity by testing around this value of N). On the other hand we have obtained $\widehat{u}_{7.5 \cdot 10^6}(3300) = 0.1879\ldots$, which yields that $\mathbb{E} \log \widetilde{Z}^{\mathrm{c}}_{3300,\omega} > 0$ with a level of error of less than $2 \cdot 10^{-9}$.

The punch-line of this discussion is that one does not need a very large

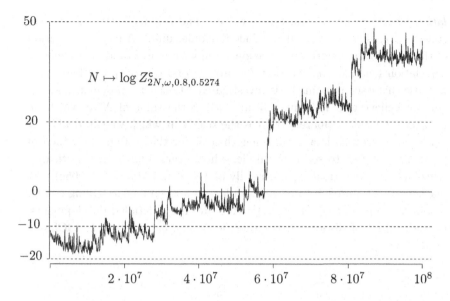

Fig. 9.1 The disordered copolymer model, based on simple random walk, with binary charges. The parameters are $\lambda = 0.8$ and $h = 0.5274$, that is about $h^{(0.825)}(0.8)$, so we are well above the lower bound on the critical curve given in Theorem 6.1. While it is difficult to estimate the free energy ($\approx 10^{-7}$?), it is *quite clear* that these data strongly suggest localization. An interesting point is definitely the very irregular growth, which proceeds by sudden bursts: this suggests that the rare stretch strategy, *cf.* Section 6.2, may be partly capturing the localization strategy close to the critical curve (the growth becomes much more regular for smaller values of h). We stress that, in order to suppress fluctuations, we have plotted only a point every 10^5 and this makes it impossible to see the fine structure of the sharp jumps. A more attentive analysis shows that these *rare regions*, that is the growth regions, are stretches of about 10^3 monomers. With reference to (9.3), we have chosen $A = 8$, $B = 3$ and $N_0 = 1000$.

system to infer localization for $\lambda = 1$ and $h = h^{(0.66667)}(1)$. But on the other hand one needs in any case a polymer of more than 3000 monomer units, which tells us that the strategy of pushing further the idea behind the argument (9.12), that is computing exactly $\mathbb{E} \log \widetilde{Z}^c_{N,\omega}$ (we are thinking of a computer assisted computation), appears to be really hopeless. For larger values of λ the size of the system can be chosen even smaller, and still observe localization, however there is a strong evidence that the minimal N is close to 1500, but not below, for λ very large (see [Caravenna *et al.* (2006)] for more details).

Another interesting aspect of the argument we have just given is that it gives us a numerical estimate of what we may call the *finite size corre-*

lation length of the system, that for $\lambda = 1$ and $h \approx 0.53$ (the case we are considering) is between 3050 and 3300 monomer units. A precise definition of this correlation length and its connection with more standard notions of correlation lengths, *e.g.* Section 7.4, are beyond our purposes here (also because this issue remains little investigated, at least for polymer models). But we believe that it is quite intuitive that the value of N at which localization becomes apparent, that is the size N at which we can *decide* for exponential growth, is a natural length-scale for the system. It is then of particular interest to see how this length diverges approaching criticality (this has been investigated numerically in [Garel and Monthus (2005c)]). A mathematical approach to correlation length and finite size scaling can be found in [Chayes *et al.* (1986)]. Further and more detailed developments, in the context of disordered Ising ferromagnets, can be found in [Chayes *et al.* (1989)].

We give in Figure 9.1 a plot of the N dependence of the logarithm of the partition function for the copolymer model. It is for a value of h corresponding to $m = 0.825$ and supposedly quite close to criticality (see Section 9.4): the highly suggestive growth pattern is discussed in the caption. Here we note in particular that for relatively small N, say up to $N = 10^7$ it appears very difficult to assess localization (one is rather tempted to bet that the system is delocalized). Precisely this suggests that deciding delocalization does not appear to be an easy matter (this point in discussed at length in [Caravenna *et al.* (2006)]).

For the disordered pinning model we have performed statistical tests in the spirit of the ones reported for copolymers. We will not report the details, but we draw the attention of the reader to Figure 9.2. This figure strongly suggests that, for $\alpha = 1/2$ and $\beta = 0.1$, the quenched critical point is very close to the anneal critical point $\log M(\beta)$. This suggests that by performing numerical tests one can improve, of course in a *statistical sense*, the bound in Theorem 5.2. And this is indeed the case, even if the numerical tests cannot be performed at the very large values of h chosen in Figure 9.2 (one can however choose $h > 0.003$, for $\lambda = 0.1$).

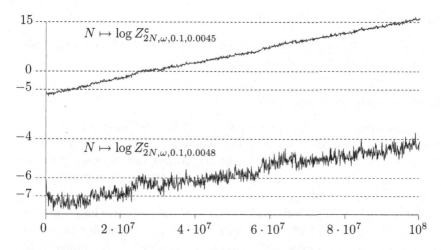

Fig. 9.2   The disordered pinning model, based on simple random walk, with binary charges. In the two cases $\beta = 0.1$, h instead takes value 0.0045 and 0.0048. In both cases to suppress fluctuations we plot only a point every 10^5. A slope is visible in both cases and it suggests a free energy of about 10^{-7} in the first case and of about $1.5 \cdot 10^{-8}$ in the second (note that the slope has to be divided by 2 since the real length of the system is $2N$). Note also that in the second case the partition function is still well below 1 for a system of length 200 millions. For $h = 0.0049$ (data not plotted) the slope is less clear. However, since $\log \cosh(0.1) = 0.004991\ldots$, these graphs suggest that the quenched critical point is very close to the annealed one (see Section 5.5). With reference to (9.3), we have chosen $A = B = 6$ and $N_0 = 1000$.

9.4   Bibliographic Complements

Complements to Section 9.1

In Section 9.1.1 and Section 9.1.2 the standard transfer-matrix approach (treated in most of the textbooks on statistical mechanics) is implemented. The Fixman–Freire algorithm is based on an idea that is quite ubiquitous in numerical analysis, *i.e.* approximating polynomial decay by superposing exponential functions, but it has been first applied in this context by [Poland (1974)] and [Fixman and Freire (1977)]. For the Fixman–Freire algorithm applied to copolymers see [Garel and Monthus (2005c)]. This algorithm is also at the base of the MELTSIM, a computer program for calculating melting (denaturation) curves and maps of DNA (the code is available from `bioinformatics.org/meltsim/`, along with a list of references to related work).

Complements to Section 9.2

This is essentially taken from [Caravenna *et al.* (2006)], where the analysis is performed with $Z_{N,\omega}^{\mathsf{c}}$ instead of $\widetilde{Z}_{N,\omega}^{\mathsf{c}}$.

Complements to Section 9.3

Much numerical work has been made on random polymers and it is impossible to account for it here, we just mention for example [Marenduzzo *et al.* (2001)], [Garel and Monthus (2005a)], [Garel and Monthus (2005c)], [Soteros and Whittington (2004)], [Causo and Whittington (2003)] and [Habibzadah *et al.* (2006)].

But a work that precisely addresses one of the issues we have raised is [Garel and Monthus (2005c)], in which results that are parallel to and in agreement with the results in [Caravenna *et al.* (2006)] are obtained. In particular both papers agree on estimating the critical value of the slope at the origin for the copolymer model between 0.82 and 0.84 (see Figure 9.1 and its caption). Such estimates however are not on particularly firm ground, in the sense that they are far from having a theoretical background comparable to that of the statistical test of Section 9.2.

We want to stress that in [Caravenna *et al.* (2006)] particular care has been put into analyzing the role of the pseudo-random number generator, above all when applying the statistical test. Several random number generators have been compared and also *true randomness* has been used (see the paper for details).

In the numerical experiments reported in this section we have used G5 CPUs clocked at 2 MHz. We have implemented the algorithms in Section 9.1.1 in C: the codes are available from the webpage of the author. The pseudo-random number generator that we have used is a standard implementation of the Mersenne–Twister algorithm [Matsumoto and Nishimura (1998)]. The figures in this section have been done with R [R Development Core Team (2004)].

Appendix A

Mathematical Tools

This Appendix is a collection of basic mathematical results, mostly in probability theory or oriented toward probability, that are repeatedly used in the body of this work.

A.1 Limit Theorems in Probability

Let $\{\xi_n\}_n$ be a sequence of Independent Identically Distributed (IID) random variables on the probability space $(\Omega, \mathcal{G}, \mathbb{P})$. The following statement is the Strong Law of Large Numbers:

Theorem A.1 *If $\xi_1 \in L^1$ then the sequence $\left\{\sum_{n=1}^{N} \xi_n / N\right\}_N$ converges, both a.s. and in the L^1 sense, to $\mathbb{E}[\xi_1]$.*

Since $\xi_{N+1}/(N+1) = \sum_{n=1}^{N+1} \xi_n/(N+1) - (N/(N+1))\sum_{n=1}^{N} \xi_n/N$, Theorem A.1 implies that if $\xi_1 \in L^1$

$$\lim_{N \to \infty} \frac{\xi_N}{N} = 0, \tag{A.1}$$

both a.s. and in the L^1 sense.

If $\{\eta_n\}_n$ is a sequence of non-negative random variable such that $\sum_n \mathbb{E}[\eta_n] < \infty$ or, equivalently (by the Fubini–Tonelli Theorem) $\mathbb{E}[\sum_n \eta_n] < \infty$, then $\sum_n \eta_n < \infty$ a.s., which in turn implies $\lim_n \eta_n = 0$ a.s.. When we apply this observation, that we will call *Borel–Cantelli argument*, to $\eta_n = \mathbf{1}_{E_n}$, $\{E_n\}_n$ a sequence of *events* (*i.e.* $E_n \in \mathcal{G}$) we obtain the first of the Borel–Cantelli lemmas:

$$\sum_n \mathbb{P}(E_n) < \infty \implies \lim_{n \to \infty} \mathbf{1}_{E_n} = 0 \text{ a.s.}, \tag{A.2}$$

which of course is equivalent to the $\mathbb{P}(d\omega)$–a.s. existence of $n(\omega) \in \mathbb{N}$ such that $\omega \notin E_n$ for every $n \geq n(\omega)$. One directly sees that the implication in (A.2) is an equivalence if $\{E_n\}_n$ is a sequence of independent events (this is the second of the Borel–Cantelli lemmas). As an immediate consequence of the formula $\sum_{n=0}^{\infty} \mathbb{P}(|\xi_1| \geq n) = \mathbb{E}[\lfloor|\xi_1|\rfloor]$ and of the second of the Borel–Cantelli lemmas one sees that $\xi_1 \in L^1$ is also necessary for (A.1). This shows also that the integrability of ξ_1 is necessary for the conclusions of Theorem A.1 to hold.

Let us recall also the Central Limit Theorem for square integrable IID sequences: if $\xi_1 \in L^2$, $\mathbb{E}[\xi_1] = 0$ and $\mathbb{E}[\xi_1^2] = 1$ then

$$\sum_{n=1}^{N} \xi_n / \sqrt{N} \stackrel{N \to \infty}{\Longrightarrow} Z \sim \mathcal{N}(0, 1). \tag{A.3}$$

Of course the symbol $\Longrightarrow$ denotes convergence in law, that we will also call weak convergence. In general we say that a sequence $\{\eta_n\}_n$, η_n a measurable application from Ω to a topological space $\mathbb{T}$ converges in law to (the measurable application) η if $\mathbb{E}[h(\eta_n)]$ converges to $\mathbb{E}[h(\eta)]$ for every bounded continuous function $h : \mathbb{T} \to \mathbb{R}$. An important example is the one in which η_n belongs to a product space (in our case it will often be $\mathbb{R}^{\mathbb{N}}$, but one can safely replace $\mathbb{N}$ with an uncountable space). In this case one directly sees hat $\eta_n \stackrel{n \to \infty}{\Longrightarrow} \eta$ if and only if for every $k \in \mathbb{N}$ the k-dimensional marginal of η_n, *i.e.* the random vector $((\eta_n)_1, \ldots, (\eta_n)_k)$, converges in law to the k-dimensional marginal of η.

A *functional* generalization of the Central Limit Theorem (A.3), called Donsker's Theorem, is precisely obtained by choosing $\mathbb{T} := C^0([0, 1]; \mathbb{R})$ (the topology is the one of uniform convergence) and by setting $B_n(t) := n^{-1/2} \sum_{j=1}^{tn} \xi_j$ for $t \in [0, 1] \cap (\mathbb{Z}/n)$. $B_n : [0, 1] \to \mathbb{R}$ is then defined by linear interpolation, so that $B_n \in C^0([0, 1]; \mathbb{R})$ (but we will also view B_n as a random *variable* taking values in $C^0([0, 1]; \mathbb{R})$). We will prefer to say that B_n is a stochastic process with trajectories in $C^0([0, 1]; \mathbb{R})$. Then under the same hypothesis as for (A.3), $B_n \stackrel{n \to \infty}{\Longrightarrow} B$, where B is the standard Brownian motion, that is the stochastic process with trajectories in $C^0([0, 1]; \mathbb{R})$ characterized by the fact that $(B_{t_1}, \ldots, B_{t_k})$ is a centered Gaussian vector, for every choice of k and $t_1, \ldots, t_k$, with $\mathbb{E}[B_{t_k} B_{t_j}] = \min(t_k, t_j)$ [Billingsley (1968)].

Another relevant improvement of (A.3) is the so called *local* central limit theorem, see *e.g.* [Petrov (1975)]. In the discrete set-up it takes the following form: if $\xi_1 \in L^2$, ξ_1 centered and of variance $\sigma^2 > 0$, takes values in $\mathbb{Z}$ and if no sub-lattice of $\mathbb{Z}$ contains the support of ξ_1 then

$$\sigma n^{1/2} \mathbf{P} \left(\frac{1}{\sigma n^{1/2}} \sum_{j=1}^{n} \xi_j = x \right) = f_Z(x) + o(1), \qquad (A.4)$$

uniformly in $x\sigma n^{1/2} \in \mathbb{Z}$. Of course $f_Z(x) = \exp(-x^2/2)/\sqrt{2\pi}$ is the density of Z. By adding conditions on ξ_1 one can go beyond and obtain a development: for example if $\xi \in L^4$ there exist two polynomials $H_1(\cdot)$ and $H_2(\cdot)$ such that

$$\sigma n^{1/2} \mathbf{P} \left(\frac{1}{\sigma n^{1/2}} \sum_{j=1}^{n} \xi_j = x \right) = f_Z(x) \left[1 + \frac{H_1(x)}{\sqrt{n}} + \frac{H_2(x)}{n} \right] + o(1/n),$$

$$(A.5)$$

still uniformly in the choice of $x\sigma n^{1/2} \in \mathbb{Z}$ [Petrov (1975), Theorem 13, Chapter VII].

Let us give some notions on the classical Cramér Large Deviations results, see [Deuschel and Stroock (1989), Chapter 1], [Dembo and Zeitouni (1998), § 2.2] or [Olivieri and Vares (2005)]. Let us set $M(\lambda) := \mathbb{E} \left[\exp(\lambda \xi_1) \right]$: it is easy to see that $M(\cdot)$ is C^∞ in the interior of the interval $D_M := \{\lambda : M(\lambda) < \infty\}$. We assume that 0 is in the interior of this set, namely that ξ_1 is locally exponentially integrable. We call $\Sigma(\cdot)$ the Legendre (or Fenchel–Legendre) transform of $\log M(\cdot)$, that is

$$\Sigma(x) := \sup_{\lambda \in \mathbb{R}} \left(\lambda x - \log M(\lambda) \right). \qquad (A.6)$$

Then $\Sigma(\cdot)$ is a lower semicontinuous (*i.e.* $\lim_n \Sigma(x_n) \geq \Sigma(x)$ for every $\{x_n\}_n$ converging to x) and convex functions taking values in $[0, \infty]$. By the Duality Lemma (see *e.g.* [Dembo and Zeitouni (1998), Lemma 4.5.8])

$$\log M(\lambda) := \sup_{x \in \mathbb{R}} \left(\lambda x - \Sigma(x) \right), \qquad (A.7)$$

at least if $\lambda \notin \partial D_M$: $M(\cdot)$ is convex but, in general, not lower semicontinuous, so what we obtain from (A.7) is a lower semicontinuous modification of $M(\cdot)$. We stress that $\Sigma(\cdot)$ is what is usually called a *good rate functional*, *i.e.* $\Sigma : \mathbb{R} \to [0, \infty]$ is lower semicontinuous and, for every $a \geq 0$, $\{x : \Sigma(x) \leq a\}$ is a compact set. Moreover $\Sigma(\cdot)$ is strictly convex and C^∞ on $\overset{\circ}{D}_M$ and $\Sigma(\lambda) = 0$ if and only if $\lambda = \mathbb{E}[\xi_1]$.

We will make use of the following version of Cramér result: for every $y \geq \mathbb{E}[\xi_1]$ we have

$$\lim_{N \to \infty} \frac{1}{N} \log \mathbb{P}\left(\frac{1}{N} \sum_{n=1}^{N} \xi_n \geq y\right) = -\Sigma(y). \qquad (A.8)$$

A.1.1   *Some remarks on convexity*

A set $D \subset \mathbb{R}^d$ is convex if $x, y \in D$ and $t \in (0, 1)$ imply $tx + (1 - t)y \in D$. A function $h : D \to \mathbb{R} \cup \{\infty\}$ is convex if for every $x, y \in D$ and $t \in (0, 1)$ we have $h(tx + (1 - t)y) \geq th(x) + (1 - t)h(y)$ (for strict convexity replace $\geq$ with $>$). These definitions extend immediately if we replace $\mathbb{R}^d$ with an arbitrary vector space. From the definition it follows immediately that if h is a convex function then $\{x : h(x) \leq a\}$ is convex for every a. In general, we may always extend the domain of h to all $\mathbb{R}^d$ by setting $h(x) = +\infty$ if $x \in \mathbb{R}^d \backslash D$. Note that if $\{h_n\}_n$ is a sequence of convex functions, $\inf_n h_n$ is a convex function and if $\lim_{n \to \infty} h_n(x) =: h(x) \in (-\infty, +\infty]$ exists for every x, then h is convex too.

If D is an open interval and $h : D \to \mathbb{R}$ is convex, then for every $x \in D$ there exists a closed interval I_x such that $h(y) - h(x) \geq c(y - x)$ for every y and $c \in I_x$. We call I_x the sub-differential of h at x. Of course h is differentiable at x if I_x contains only one point, $h'(x)$, and it is immediate to see that h is differentiable outside of a countable set. If $\{h_n\}_n$ is a sequence of differentiable convex functions converging pointwise to h, then for x in the interior of $\{y : h(y) < \infty\}$ we have that both the inferior and the superior limit of $\{h_n'(x)\}_n$ are in the sub-differential I_x of h at x. In particular, if h is differentiable at x then $\lim_{n \to \infty} h_n'(x) = h'(x)$.

All these results are easily proven by using the monotonicity properties of incremental ratios $\Delta_h(x, y) := (h(y) - h(x))/(y - x)$. In particular, by setting $\partial_- h(x) := \lim_{y \nearrow x} \Delta_h(x, y)$ and $\partial_+ h(x) := \lim_{y \searrow x} \Delta_h(x, y)$, one obtains $I_x = [\partial_- h(x), \partial_+ h(x)]$.

A.2   Entropy

If ν and μ are two probability measures on the same measurable space and if $\nu \ll \mu$ (in words: ν is absolutely continuous with respect to μ, that is if

A is measurable and $\mu(A) = 0$, then $\nu(A) = 0$ too) we set

$$\mathcal{S}(\nu|\mu) := \int \log\left(\frac{d\nu}{d\mu}\right) d\nu. \tag{A.9}$$

$\mathcal{S}(\nu|\mu)$ is the *relative entropy*, or *Kullback information*, of ν with respect to μ. If $\nu \not\ll \mu$ then we set $\mathcal{S}(\nu|\mu) := \infty$.

Let us assume $X \in L^1(\nu)$ and that both $\mu \ll \nu$ and $\nu \ll \mu$ (in this case we say that μ and ν are equivalent). Then the following inequality holds:

$$\log \int \exp(X) \, d\mu \geq \int X \, d\nu - \mathcal{S}(\nu|\mu). \tag{A.10}$$

We may assume that $\int \exp(X) \, d\mu < \infty$, otherwise there is nothing to prove. By Jensen inequality we have

$$\log \int \exp(X) \, d\mu = \log \int \exp\left(X - \log\left(\frac{d\nu}{d\mu}\right)\right) d\nu$$
$$\geq \int X \, d\nu - \int \log\left(\frac{d\nu}{d\mu}\right) d\nu. \tag{A.11}$$

We recognize $\mathcal{S}(\nu|\mu)$ in the last term and (A.10) is proven.

Inequality (A.10) is often extremely helpful in making precise the so called *energy-entropy* arguments. Interpret X as the energy: $\int \exp(X) \, d\mu$ should depend heavily on the large values of X, but this may not be true if these large values are taken up in a region of the probability space that is weighted little by μ. There is therefore a competition between the energy and the way in which the probability mass is distributed, that we call informally *entropic contribution*. From a technical viewpoint, using inequality (A.10) requires guessing a measure ν for which we can explicitly compute or estimate $\int X \, d\nu$ and $\mathcal{S}(\nu|\mu)$ and that, at the same time, yields overall an interesting estimate. In this sense the following variational formula holds guaranties that our procedure is, in principle, optimal:

$$\log \int \exp(X) \, d\mu = \sup_\nu \left(\int X \, d\nu - \mathcal{S}(\nu|\mu)\right), \tag{A.12}$$

where the supremum is taken over probabilities ν which are equivalent to μ. If $\int \exp(X) \, d\mu < \infty$ the formula follows from (A.10) and the choice $d\nu = \exp(X) \, d\mu / \int \exp(X) \, d\mu$. If $\int \exp(X) \, d\mu = +\infty$ just choose $d\nu = \exp(X_n) \, d\mu / \int \exp(X_n) \, d\mu$, with $X_n = \min(X, n)$ and then let $n \to \infty$.

In estimating probabilities from below we use an entropy inequality that we are going to introduce now. Let once again ν and μ be two equivalent probability measures. If E is an event such that $\nu(E) > 0$ then

$$\log\left(\frac{\mu(E)}{\nu(E)}\right) \geq -\frac{1}{\nu(E)}\left(\mathcal{S}(\nu|\mu) + e^{-1}\right). \tag{A.13}$$

The application of this inequality for Large Deviations lower bounds is immediate: if $\{E_n\}_n$ is such that $\mu(E_n) \overset{n\to\infty}{\to} 0$, then if one can find $\{\nu_n\}_n$ that $\nu_n(E_n) \overset{n\to\infty}{\to} 1$, then $\log\mu(E_n) \geq -\mathcal{S}(\nu_n|\mu) + O(1)$.

The inequality (A.13) follows by observing that

$$\log\left(\frac{\mu(E)}{\nu(E)}\right) = \log\int_E \frac{\mathrm{d}\mu}{\mathrm{d}\nu}\frac{\mathrm{d}\nu}{\nu(E)} \geq \int_E \log\left(\frac{\mathrm{d}\mu}{\mathrm{d}\nu}\right)\frac{\mathrm{d}\nu}{\nu(E)}$$
$$= -\frac{1}{\nu(E)}\int_E \left(\frac{\mathrm{d}\nu}{\mathrm{d}\mu}\log\frac{\mathrm{d}\nu}{\mathrm{d}\mu}\right)\mathrm{d}\mu \geq -\frac{1}{\nu(E)}\left(\int\left(\frac{\mathrm{d}\nu}{\mathrm{d}\mu}\log\frac{\mathrm{d}\nu}{\mathrm{d}\mu}\right) + \frac{1}{e}\right), \tag{A.14}$$

where in the first inequality we have applied Jensen's inequality and in the second inequality we have used that $\min_{r>0} r\log r = -1/e$.

A.3 Concentration Properties

A number of inequalities on the so called *concentration of measure phenomenon* ([Talagrand (1996)], [Ledoux (2001)], [Villani (2003)]) are available for product probability spaces. We say that the IID sequence ξ of integrable random variables satisfies a concentration inequality if there exists a continuous function $\vartheta : [0,\infty) \to [0,\infty)$, $\lim_{t\to\infty}\vartheta(t) = 0$, such that for every n and every Lipschitz convex function $G : \mathbb{R}^n \to \mathbb{R}$ with Lipschitz constant 1, that is

$$|G(x_1,\ldots,x_n) - G(y_1,\ldots,y_n)| \leq |x - y| = \sqrt{\sum_{j=1}^n (x_j - y_j)^2}, \tag{A.15}$$

for every x and $y \in \mathbb{R}^n$, we have

$$\mathbb{P}\left(|G(\xi_1,\ldots,\xi_n) - \mathbb{E}G(\xi_1,\ldots,\xi_n)| \geq t\right) \leq \vartheta(t), \tag{A.16}$$

for every $t \geq 0$. The convexity assumption is actually required only for treating the case of general bounded charges, see (1) below, and in all the other cases it can be removed: however in all the applications convexity is verified.

We now give a list of cases in which (A.16), specifying of course the function $\vartheta(\cdot)$.

(1) For any bounded random variable, that is when there exists a constant $c > 0$ such that $\mathbb{P}(|\xi_1| > c) = 0$, (A.16) holds with $\vartheta(t) = c_1 \exp(-c_2 t^2)$ for suitable positive constants c_1 and c_2, depending only on c, under the further hypothesis that $G(\cdot)$ is convex. This result may be found for example in [Ledoux (2001), Corollary 4.10 and Proposition 1.8].

(2) The inequality (A.16) holds with $\vartheta(t) = c_1 \exp(-c_2 t^2)$ for any random variable ξ_1 whose law satisfies a logarithmic Sobolev inequality [Ledoux (2001), Chapter 5]. The logarithmic Sobolev inequalities are a family of inequalities depending on a constant, the *log-Sobolev constant*, The constant c_2 depends on the log-Sobolev constant. We will not go into the details of such a class of random variables, but we simply stress that if ξ_1 is a continuous random variable with density $f_{\xi_1}(\cdot) =: \exp(-V(\cdot))$ such that $V(\cdot) \in C^2$ and $V(\cdot)$ is strictly convex on (K, ∞) and on $(-\infty, -K)$ for some $K > 0$, then the law of ξ_1 satisfies a logarithmic Sobolev inequality (see [Ané *et al.* (2000), Chapter 6 and Theorem 3.2.2] for a more general condition and for a proof of this fact). Of course $\xi_1 \sim \mathcal{N}(0,1)$ is in this class of variables: more generally, if $V(t) = const.t^p$, ≥ 2, then ξ_1 satisfies a logarithmic Sobolev inequality.

(3) The inequality (A.16) holds with $\vartheta(t) = c_1 \exp(-c_2 t)$ for any random variable ξ_1 whose law satisfies a Poincaré inequality, [Ledoux (2001), Chapter 3] and [Ané *et al.* (2000), Chapter 6 and Theorem 3.2.1]. In the latter reference one can find also explicit conditions on $f_{\xi_1}(\cdot) =: \exp(-V(\cdot))$ for the Poincaré inequality to hold. The constant c_2 depends on the Poincaré constant, or spectral gap, appearing in the inequality. Examples of cases in which the Poincaré inequality holds include

- $V(\cdot) \in C^2$ with $\sup_t |V''(t)| < \infty$ and $\inf_{|t|>a} V'(t) > 0$ for some $a > 0$;
- $V(t) = const.t^p$, $p \geq 1$.

Much effort has been put into improving the constant c_1 and c_2 in this (and similar) inequalities, for particular classes of functions, in particular the $\mathcal{N}(0,1)$ case. In our case explicit constants are relevant only at one instance (in Chapter 9). In that case we use [Ané *et al.* (2000), Theorem 7.4.4] which states that if $\mathbb{P}(\omega_1 = +1) = \mathbb{P}(\omega_1 = -1) = 1/2$ and if $G : \{-1, +1\}^n \to \mathbb{R}$ is such that $\sup_x \sum_{i=1}^n (G(\theta_i x) - G(x))^2 \leq 1$, with

$(\theta_i x)_i = -x_i$ and $(\theta_i x)_j = x_j$ for $j \neq i$, then

$$\mathbb{P}\left(G(\omega) - \mathbb{E}G(\omega) \geq t\right) \leq \exp(-t^2), \tag{A.17}$$

for every $t \geq 0$. We observe that if $G : [-1,1]^n \rightarrow \mathbb{R}$ is differentiable and if $\sup_{i,x} |\partial_{x_i} G(x)| \leq c$, then restricting $G(\cdot)$ to $\{-1,1\}^n$ we have $\sup_x \sum_{i=1}^n (G(\theta_i x) - G(x))^2 \leq 4c^2 n$.

A.4   Slow Variation and Laplace Transforms

A very complete reference for the content of this section is [Bingham *et al.* (1987)], but all the results we state may be found also in [Feller (1971)]. A measurable function $L : (0, \infty) \rightarrow (0, \infty)$ is slowly varying at infinity if $\lim_{x \to \infty} L(\kappa x)/L(x) = 1$ for every $\kappa > 0$. It can be shown, [Bingham *et al.* (1987), Theorem 1.2.1], that such a convergence holds uniformly in κ, whenever κ belongs to a bounded interval of $(0, \infty)$, namely: for every $0 < c_1 < c_2 < \infty$

$$\lim_{x \to \infty} \sup_{\kappa \in [c_1, c_2]} \left| \frac{L(\kappa x)}{L(x)} - 1 \right| = 0. \tag{A.18}$$

Examples of slowly varying functions include $(\log(1 + x))^c$, $c \in \mathbb{R}$ or any positive measurable function that has a positive limit at infinity (and in this case we speak of *trivial* slowly varying function). However the class is richer and it contains *non-logarithmic* functions like $\exp((\log(x + 2))^\beta)$, $\beta \in (0, 1)$. They can actually be completely characterized [Bingham *et al.* (1987), Theorem 1.3.1] in terms of two bounded measurable functions c and $\varepsilon : (0, \infty) \rightarrow \mathbb{R}$ such that $\lim_{x \to \infty} c(x) \in \mathbb{R}$ and $\lim_{x \to \infty} \varepsilon(x) = 0$: for $x \geq 1$

$$L(x) = \exp\left(c(x) + \int_1^x \frac{\varepsilon(t)}{t} \, dt \right). \tag{A.19}$$

If we say that $L : \mathbb{N} \rightarrow (0, \infty)$ is slowly varying we mean that $x \mapsto L(\lceil x \rceil)$ is slowly varying in the standard sense and we will implicitly extend the domain of $L(\cdot)$ to $(0, \infty)$.

We collect here a number of useful results on the (generic) slowly varying function $L(\cdot)$:

L.1 For every $\epsilon > 0$ we have

$$\lim_{x \to \infty} x^\epsilon L(x) = +\infty \quad \text{and} \quad \lim_{x \to \infty} x^{-\epsilon} L(x) = 0. \tag{A.20}$$

See [Bingham *et al.* (1987), Proposition 1.3.6].

L.2 For every $\beta > -1$

$$\sum_{n=1}^{N} n^\beta L(n) \stackrel{N\to\infty}{\sim} \frac{N^{\beta+1}}{\beta+1} L(N). \tag{A.21}$$

See [Bingham *et al.* (1987), Proposition 1.5.8] by keeping into account that it is straightforward, by using L.1, to see $\sum_{n=1}^{N} L(n)n^\beta \sim \int_1^N L(x)x^\beta \, \mathrm{d}x$ as $N \to \infty$ (analogous observations will be repeatedly used).

L.3 For every $\beta < -1$

$$\sum_{n \geq N} n^\beta L(n) \stackrel{N\to\infty}{\sim} \frac{N^{\beta+1}}{-(\beta+1)} L(N). \tag{A.22}$$

See [Bingham *et al.* (1987), Proposition 1.5.10].

L.4 If $\beta = -1$ we will use the following: if $\sum_n L(n)/n = \infty$ then there exists a slowly varying function $\hat{L}(\cdot)$ such that $\hat{L}(N) = \sum_{n=1}^{N} L(n)/n$. Moreover $\lim_{N\to\infty} \hat{L}(N)/L(N) = \infty$. If instead $\sum_n L(n)/n < \infty$, then there exists a slowly varying function $\widetilde{L}(\cdot)$ such that $\widetilde{L}(N) = \sum_{n=N+1}^{\infty} L(n)/n$ and $\lim_{N\to\infty} \widetilde{L}(N)/L(N) = \infty$ [Bingham *et al.* (1987), Proposition 1.5.9a and b].

L.5 If $\beta > 0$ and if $u(b) := b^\beta L(1/b)$ then $u(\cdot)$ is increasing in $(0, \delta)$ for some $\delta > 0$ and $u^{-1}(x) = x^{1/\beta} L_\beta(x)$, with $L_\beta(x)$ a slowly varying function of x. The analogous result for $\beta < 0$ holds, but this time $u(\cdot)$ is (eventually) decreasing and $L_\beta(x)$ has to be replaced by $L_\beta(1/x)$. See [Bingham *et al.* (1987), Theorem 1.5.12].

In the standard terminology $L_0(\cdot)$ is slowly varying at zero if $L_0(b) := L(1/b)$ for some function $L(\cdot)$ which is slowly varying at infinity. Moreover the function $x \to x^\beta L(x)$ is usually called *regularly varying* (at infinity). However for us *slowly varying* means *slowly varying at infinity* and we will avoid talking explicitly of regular variation.

We use the standard definition $\Gamma(\alpha) := \int_0^\infty t^{\alpha-1} \exp(-t) \, \mathrm{d}t$. $\Gamma(\alpha)$ is defined for every $\alpha \in [0, \infty)$. One readily verifies that $\Gamma(\alpha + 1) = \alpha\Gamma(\alpha)$ for every $\alpha > 0$ and that $\Gamma(1) = 1$, so that $\Gamma(n+1) = n!$, $n \in \mathbb{N} \cup \{0\}$. The same fact yields also $\Gamma(\alpha) \sim 1/\alpha$ for $\alpha \searrow 0$. Just a bit harder to verify is that $\Gamma(\alpha)\Gamma(1-\alpha) = \pi/\sin(\pi\alpha)$. We will make use of

Theorem A.2 *Let $u : \mathbb{N} \to [0, \infty)$. If $\sum_{n=1}^{N} u(n) \overset{N \to \infty}{\sim} N^{\beta} L(N)$, with $\beta \geq 0$, then*

$$\sum_{n=1}^{\infty} \exp(-bn)u(n) \overset{b \searrow 0}{\sim} \Gamma(1+\beta)\frac{L(1/b)}{b^{\beta}}. \tag{A.23}$$

Moreover if $\sum_{n} u(n) = 1$, that is $u(\cdot)$ is a discrete probability density, then if $\sum_{n>N} u(n) \overset{N \to \infty}{\sim} L(N)/N^{\beta}$, with $\beta \in [0, 1)$, we have

$$1 - \sum_{n=1}^{\infty} \exp(-bn)u(n) \overset{b \searrow 0}{\sim} \Gamma(1-\beta)b^{\beta}L(1/\beta). \tag{A.24}$$

For a proof of Theorem A.2 see [Bingham *et al.* (1987), Theorem 1.7.1 and Corollary 8.1.7]. We stress however that the two theorems to which we have just referred, called *Tauberian* theorem, give also the converse implications and *Tauberian* precisely refers to the converse statements. Theorem A.2 just gives the asymptotic behavior of the Laplace transform, given the asymptotic behavior of the function and it is appropriate to refer to it as an *Abelian* theorem.

It may be useful to stress that, if $\beta \in (0, 1)$, (A.24) can be derived from (A.23) as a direct application of the summation by parts formula

$$1 - \sum_{n=1}^{\infty} \exp(-bn)u(n) = (1 - \exp(-b)) \sum_{n=0}^{\infty} \exp(-bn) \sum_{j>n} u(j), \tag{A.25}$$

and L.2.

A.5 Renewal Theory for $\mathbb{N}$ Valued Random Variables

Key reference for this section is [Asmussen (2003)]. Given the discrete probability density $K(\cdot)$ on $\mathbb{N}$, $\tau := \{\tau_j\}_{j=0,1,\ldots}$ is the sequence of partial sums of an IID sequence of variables distributed according to $K(\cdot)$, that is $\tau_0 = 0$ and the sequence of *inter-arrival times* $\{\tau_j - \tau_{j-1}\}_{j=1,2,\ldots}$ is IID, $\tau_1 \sim K(\cdot)$. We call τ *renewal process* on $\mathbb{N}$ associated to $K(\cdot)$. τ may be also viewed as a point process, that is as a random subset of $\mathbb{N} \cup \{0\}$, so, for example, $n \in \tau$ means that there exists j such that $\tau_j = n$. Actually, the points in τ will be called *renewals* or *renewal points*. If $\widetilde{\tau}_0$ is a $\mathbb{N}$–valued random variable independent of τ, then we call $\widetilde{\tau}_0 + \tau$ *delayed renewal process* and $\widetilde{\tau}_0$ is the *delay*. Delayed or not, renewal processes enjoy the renewal property, *i.e.* if E is in the σ-algebra generated by the events $\{A \subset \tilde{\tau}_0 + \tau\}$, with $A \subset \{1, \ldots, n\}$, and if F is in the σ-algebra generated by $\{B \subset \tilde{\tau}_0 + \tau\}$, with $B \subset \{n+1, n+2, \ldots\}$, then $\mathbf{P}\left(E \cap \{n \in \tilde{\tau}_0 + \tau\} \cap F\right)$ is equal to $\mathbf{P}\left(E \cap \{n \in \tilde{\tau}_0 + \tau\}\right) \mathbf{P}\left(\tau + n \in F\right)$.

In general one may replace $\mathbb{N}$ with $\mathbb{N} \cup \{\infty\}$, by setting $K(\infty) = 1 - \Sigma_K$, $\Sigma_K = \sum_{n \in \mathbb{N}} K(n)$. Of course this is useless if $K(\cdot)$ is a probability, but this is no longer the case if $\Sigma_K < 1$: in this case we will say that $K(\cdot)$ is defective and the associated renewal process is *terminating*, in the sense that a.s. $\tau_j = \infty$ for j sufficiently large. Probabilities and defective probabilities will be generically named sub-probabilities.

Renewal processes are naturally linked to Markov chains, we suggest [Asmussen (2003), Chapter I] for a quick review on Markov chains and for all basic notions. By the strong Markov property the sequence of successive returns of a Markov chain to a fixed (recurrent or transient) state is a renewal process. Moreover this picture is general: any renewal process can be constructed as return time of a suitable Markov chain. In fact if we define $A_n := A_n(\tau) := n - \sup\{\tau_k : \tau_k \leq n\}$, then the sequence $A := \{A_n\}_{n=0,1,\ldots}$ is a Markov chain called *backward recurrence time*. Note that $A_n \in \mathbb{N} \cup \{0\}$ is the time passed since the last renewal when looking from n, see Figure A.1. The probability transition from $A_n = i$ to $A_{n+1} = j$ is non-zero only if $j = i + 1$ or $j = 0$ and the probability that the process moves up is $\overline{K}(i+1) + K(\infty) / (\overline{K}(i) + K(\infty))$, independently of $A_0, A_1, \ldots, A_{n-1}$ (we have used $\overline{K}(i) := \sum_{n > i} K(n)$, but in this sum $K(\infty)$ is not included).

We say that $K(\cdot)$ has period $p \in \mathbb{N}$ if $\{n : K(n) > 0\}$ is contained in $\{pn : n \in \mathbb{N}\}$ and if p is the largest number with this property. If $p = 1$ we say that $K(\cdot)$ is aperiodic. The aperiodicity of $K(\cdot)$ implies the aperiodicity of the Markov chain A. Note that when $\Sigma = 1$, the state space of A is $\{0, 1, \ldots, \sup\{n : K(n) > 0\}\}$ and that A is irreducible (that is all states communicate in a finite number of steps, with positive probability). Moreover if $\Sigma_K < 1$, that is when τ is terminating, the state space may be decomposed into the transient class $\mathbb{N} \cup \{0\}$ and the recurrent class $\{\infty\}$.

We set $m_K := \sum_{n \in \mathbb{N} \cup \{\infty\}} nK(n) \in [1, \infty]$. Note that $m_K = \infty$ may arise also when $K(\infty) = 0$: in this case A is a recurrent Markov chain, but it is immediate to see that it is a null recurrent chain, since τ_1 coincides with $\inf\{n > 0 : A_n = 0\}$. On the other hand, A is clearly positive recurrent if $m_K < \infty$. We will therefore say that τ is positive (respectively null) recurrent if A is. We will therefore also say that τ is transient when it is terminating.

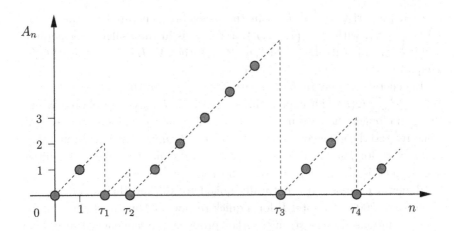

Fig. A.1 The A process (large gray dots) associated to the renewal τ. In this case $\tau_1 = 2$, $\tau_2 = 3$, $\tau_3 = 9$, $\tau_4 = 12$ and $\tau_5 > 13$.

A.5.1   *The renewal theorem*

The following is the Classical Renewal Theorem, see [Asmussen (2003), Chapter I, Theorem 2.2]:

Theorem A.3   *If $K(\cdot)$ is aperiodic then $\lim_{n \to \infty} \mathbf{P}\,(n \in \tau) = 1/m_K$, with $1/\infty = 0$.*

The result of course extends immediately to delayed renewal processes. Note that $\mathbf{P}\,(n \in \tau)$ may be looked upon as a *Green function*. It is in fact the expected number of times that the partial sum process τ spends in n: but of course such a process touches n at most once and the Green's function in this case is *a probability*. We will call it (also) *renewal mass function*. A useful general formula for the Green's (or renewal mass) function is

$$\mathbf{P}\,(n \in \tau) \;=\; \sum_{j=0}^{\infty} K^{j*}(n), \tag{A.26}$$

where $K^{j*}(\cdot)$ is the j–fold convolution of $K(\cdot)$, $K^{2*}(n) = K * K(n) = \sum_j K(n-j)K(j)$ and $K^{0*}(n) := \mathbf{1}_{n=0}$. Of course, for every n, the series in (A.26) is in reality a finite sum. Formula (A.26) follows from the *renewal* *equation*

$$\mathbf{P}(n \in \tau) = \sum_{k=0}^{n} \mathbf{P}(k \in \tau)K(n-k) + \mathbf{1}_{n=0}, \qquad (A.27)$$

which, in turn, is an immediate consequence of the renewal property of τ.

It is useful to note that Theorem A.3 may be proven by looking at the ergodic properties of the backward recurrence time process A [Asmussen (2003), § VII.2]. In particular one sees that the condition $m_K < \infty$ is precisely the condition for A to be positive recurrent. This implies directly the existence of a unique invariant probability measure. One can write it explicitly:

$$p_A(n) = \frac{1}{m_K} \sum_{j \geq n+1} K(j), \quad n = 0, 1, \ldots. \qquad (A.28)$$

By the Ergodic Theorem for irreducible aperiodic Markov chains one also has that

$$\lim_{N \to \infty} \mathbf{P}\left(A_N = n\right) = p_A(n). \qquad (A.29)$$

If $A = \{A_n\}_n$ is redefined so that A_0 is distributed according to $p_A(\cdot)$, so that A is stationary, then $\{n : A_n = 0\}$ is a delayed renewal that we call stationary renewal.

Associated to the renewal process τ there is also the process $\{\mathcal{N}_n(\tau)\}_{n=1,\ldots}$ defined by

$$\mathcal{N}_n(\tau) = |\tau \cap \{1, \ldots, n\}|. \qquad (A.30)$$

Note that $\mathcal{N}_n(\tau) = \sup\{j : \tau_j \leq n\}$. A direct application of the law of large numbers yields

$$\lim_{n \to \infty} \frac{1}{n}\mathcal{N}_n(\tau) = \frac{1}{m_K}, \qquad (A.31)$$

both in the almost sure and in the L^1 sense (as a matter of fact, in L^p for any $p \geq 1$).

Theorem A.3 says nothing about the speed of convergence to zero of $\mathbf{P}\left(n \in \tau\right)$ when $m_K = \infty$. We treat this case by considering separately the null recurrent ($K(\infty) = 0$) and the transient case ($K(\infty) > 0$). From now on in this section we will restrict to $K(\cdot)$ of the form given in Definition 1.4.

A.5.2   *The renewal mass function: Transient case*

In the transient case we prove the following:

Theorem A.4   *If $K(\infty) > 0$ then*

$$\mathbf{P}\left(N \in \tau\right) \overset{N \to \infty}{\sim} K(N)\frac{1}{(1-\delta)^2}, \tag{A.32}$$

with $\delta = \sum_{n \in \mathbb{N}} K(n)(< 1)$.

For the proof we introduce the discrete probability density $q : \mathbb{N} \to (0,1)$, $q(n) := K(n)/\delta$. We also introduce $\widetilde{L}(\cdot) := L(\cdot)/\delta$, so $q(n) = \widetilde{L}(n)/n^{1+\alpha}$.

We have the following estimates:

Lemma A.5   *For every k*

$$q^{k*}(n) \sim kq(n), \tag{A.33}$$

as $n \to \infty$. Moreover there exists $c > 0$ such that

$$q^{k*}(n) \le k^c q(n), \tag{A.34}$$

for every $k \in \mathbb{N}$ and for every n.

Proof of Theorem A.4. By (A.26) we have

$$\mathbf{P}\left(N \in \tau\right) = \sum_{k=0}^{\infty} \delta^k q^{k*}(N). \tag{A.35}$$

We observe that Lemma A.5 provides precisely the estimates to apply the Dominated Convergence Theorem in the formula that follows:

$$\frac{\mathbf{P}\left(N \in \tau\right)}{q(N)} = \sum_{k=0}^{\infty} \delta^k \frac{q^{k*}(N)}{q(N)} \overset{N \to \infty}{\sim} \sum_{k=1}^{\infty} k\delta^k = \frac{\delta}{(1-\delta)^2}, \tag{A.36}$$

Theorem A.4 is therefore proven.

Theorem A.4
$\square$

Proof of Lemma A.5. Equation (A.33) is manifestly true for $k = 1$. Let us suppose that it holds for $k = 1, 2, \ldots, m$. Then let us write

$$
\frac{q^{(m+1)*}(n)}{q(n)} = \left(\sum_{j=1}^{\lfloor n/2 \rfloor} + \sum_{j=\lfloor n/2 \rfloor+1}^{n-1} \right) q^{m*}(j) \frac{q(n-j)}{q(n)}
$$

$$
= \sum_{j=1}^{\lfloor n/2 \rfloor} q^{m*}(j) \frac{q(n-j)}{q(n)} + \sum_{j=1}^{\lceil n/2 \rceil-1} q(j) \frac{q^{m*}(n-j)}{q(n)}. \tag{A.37}
$$

By assumption, for any fixed j and $n \to \infty$ the two ratios in the last term converge respectively to 1 and to m and they are uniformly bounded. Then (A.33) follows as an application of the Dominated Converge Theorem.

For the proof of the (A.34) we start by observing for $k = 1$ there is nothing to prove. Let us assume that (A.34) holds for every $k < 2m$ (and every n). Then

$$
q^{2m*}(n) \le 2 \sum_{j=1}^{\lfloor n/2 \rfloor} q^{m*}(j) q^{m*}(n-j) \le 2m^c \sum_{j=1}^{\lfloor n/2 \rfloor} q^{m*}(j) q(n-j). \tag{A.38}
$$

We observe now that, by (A.18), given the slowly varying function $\widetilde{L}(\cdot)$, there exists $c_1 > 0$ such that $\widetilde{L}(xr) \le c_1 \widetilde{L}(r)$ for every $x \in [1/2, 1]$ and every $r \ge 1$. Therefore $q(n-j) = \widetilde{L}(n-j)/(n-j)^{1+\alpha} \le c_1 2^{1+\alpha} q(n)$ and

$$
q^{2m*}(n) \le c_1 2^{2+\alpha-c} (2m)^c q(n) \sum_j q^{m*}(j), \tag{A.39}
$$

and if $c = \alpha + 2 + \log_2 c_1$ then we obtain (A.34) for $k = 2m$. If instead $k = 2m+1$, we assume (A.34) for $k < 2m+1$ and by repeating (essentially) the same argument we obtain (A.34) for $k = 2m + 1$, with the same choice of c.

Lemma A.5
□

A.5.3 *The renewal mass function: Null recurrent case*

The null recurrent case is the most delicate and we will mostly refer to the literature for the proofs. We choose $K(\cdot)$ as in Definition 1.4 (and then $\alpha \in [0, 1]$). The basic results, for which we cite [Port (1963)] and [Garsia and Lamperti (1963)], are by now classical, but the theory is still actively evolving (we mention in particular [Doney (1997)]).

The first result we mention is the following [Bingham *et al.* (1987), Theorem 8.7.3 and Theorem 8.7.5]:

Theorem A.6 *For $\alpha \in (0,1)$ and $K(\infty) = 0$ we have that*

$$\mathbf{E}\left[\mathcal{N}_N(\tau)\right] = \sum_{n=1}^{N} \mathbf{P}(n \in \tau) \overset{N \to \infty}{\sim} \frac{\sin(\pi\alpha)}{\pi} \frac{N^\alpha}{L(N)}. \tag{A.40}$$

If $\alpha = 0$ (A.40) still holds, but $L(\cdot)$ has to be replaced with a different slowly varying function (and the pre-factor has to be omitted), see L.4 of Appendix A.4. Moreover for $\alpha = 1$ we have that

$$\mathbf{P}(N \in \tau) \overset{N \to \infty}{\sim} 1/\sum_{n=0}^{N} \overline{K}(n), \tag{A.41}$$

and, in particular, $N \mapsto \mathbf{P}(N \in \tau)$ is slowly varying. We remark also that, by L.3 of Appendix A.4, (A.41) implies $\sum_{n=1}^{N} \mathbf{P}(n \in \tau) \sim N/\sum_{n=0}^{N} \overline{K}(n)$.

It is worthwhile mentioning that from (A.27) one readily obtains for $b > 0$

$$\sum_n \exp(-bn)\mathbf{P}(n \in \tau) = \left(1 - \sum_n \exp(-bn)K(n)\right)^{-1}, \tag{A.42}$$

where the sum is over $\mathbb{N} \cup \{0\}$. It is therefore clear that we can easily obtain the the asymptotic behavior of the Laplace transform of the mass renewal function of τ, that is the left-hand side, and (A.40) follows by a Tauberian argument.

Note that the result for $\alpha = 1$ is sensibly stronger than the one for $\alpha < 1$. In reality, recently also for the case $\alpha \in (0,1)$ the following sharper result has been proven:

Theorem A.7 *[Doney (1997), Theorem B]. For $\alpha \in (0,1)$ and $K(\infty) = 0$ we have*

$$\mathbf{P}(N \in \tau) \overset{N \to \infty}{\sim} \frac{\alpha \sin(\pi\alpha)}{\pi} \frac{1}{L(N)N^{1-\alpha}}. \tag{A.43}$$

A.5.4 *Scaling limits of renewal processes*

Given a renewal process τ, one can consider the random subset $\tau/N \subset [0,\infty)$. Let us stress immediately that a realization of τ/N is a closed subset of $[0,\infty)$ (we could alternatively work on $[0,\infty]$ with the standard compactification of the Euclidean metric and consider instead $\tau/N \cup \{\infty\}$ when τ is recurrent, but there is no difference). Of course if the renewal is positive recurrent intuitively the limit is $[0, \infty)$ itself and if the renewal is terminating then the limit is just $\{0\}$. And in the null recurrent case a more complicate scenario comes up: think for example to the returns to zero of a simple random walk, that converge, in some sense to be specified, to the set of zeros of a Brownian motion.

In order to properly attack this problem we have to start by giving a notion of convergence for sequences of subsets of $[0, \infty)$. We will work with the class $\mathcal{C}_\infty$ of closed subsets of $[0, \infty)$ and endow it with the Matheron topology [Matheron (1975)] that we describe next. For $F \in \mathcal{C}_\infty$ and $t \geq 0$ we set $d_t(F) = \inf(F \cap (t, \infty))$. The function $t \mapsto d_t(F)$ is right-continuous and $F = \{t \geq 0 : \lim_{s \nearrow t} d_s(F) = t\}$, so F is identified by $t \mapsto d_t(F)$. The space of such functions may be endowed with the Skorohod topology on *cadlag* functions taking values on $\overline{[0, \infty)} = [0, \infty]$ and this topology gives the Matheron topology on $\mathcal{C}_\infty$ by the identification we have pointed out above. Under this topology the space $\mathcal{C}_\infty$ is metrizable, separable and compact, hence it is a Polish space. The Borel σ–algebra on $\mathcal{C}_\infty$, denoted by $\mathcal{B}(\mathcal{C}_\infty)$, coincides with the σ–algebra generated by the family of maps $\{d_t(\cdot), \ t \in [0, \infty)\}$. In particular the compactness of the Polish space $\mathcal{C}_\infty$ implies immediately that if we have a sequence of random sets whose laws are $\{\mathbf{P}_N\}_N$, this sequence is tight and it is sufficient to study the convergence of finite dimensional marginals. Namely $\mathbf{P}_N \overset{N \to \infty}{\Longrightarrow} \mathbf{P}$ if for every n and every $\underline{t} = (t_1, \ldots, t_n) \in [0, \infty)$ with $t_1 < \ldots < t_n$ one has

$$\mathbf{P}_N G_{\underline{t}}^{-1} \overset{N \to \infty}{\Longrightarrow} \mathbf{P} G_{\underline{t}}^{-1}, \tag{A.44}$$

with $G_{\underline{t}} : \mathcal{C}_\infty \to [0, \infty]^n$ given by $G_{\underline{t}}(F) = (d_{t_1}(F), \ldots, d_{t_n}(F))$.

It is relevant to remark that if the sequence $\{F_n\}_n$ of elements of $\mathcal{C}_\infty$ converges, then also $\{F_n \cap [0, t]\}_n$, $t > 0$, converges. A metric on $\mathcal{C}_1$, the closed subsets of $[0, 1]$, is the Hausdorff one:

$$\rho(F_1, F_2) := \max_{i \in \{(1,2),(2,1)\}} \sup_{t \in F_{i_1}} \inf_{s \in F_{i_2}} |t - s|, \qquad F_1, F_2 \in \mathcal{C}_1. \tag{A.45}$$

This metric is equivalent to the Matheron one (restricted to $\mathcal{C}_1$). We point out that it is also possible to define the Hausdorff metric $\mathcal{C}_\infty$ by adding the point $\{\infty\}$ to the set and by performing the standard compactification procedure.

Let us observe now that if we have a renewal process τ and, on the same probability space, a Poisson process $\{N_\gamma(t)\}_{t \geq 0}$ of intensity $\gamma > 0$ independent of the renewal, then it is easy to see that $\{\tau_{N_\gamma(t)}\}_{t \geq 0}$ is a non-decreasing process with independent stationary increments whose trajectories are right-continuous. In other words $\{\tau_{N_\gamma(t)}\}_{t\geq 0}$ is a *subordinator* [Bertoin (1996)]. By the general theory of subordinators, the law $\mathbf{P}$ of a general subordinator σ can be characterized by giving the Laplace transform of its one point marginals:

$$\mathbf{E}\left[\exp\left(-\lambda\sigma_t\right)\right] =: \exp\left(-t\Phi(\lambda)\right), \qquad (A.46)$$

for t and λ non-negative. The right-hand side defines $\Phi : [0,\infty) \to [0,\infty)$ which is called the *Lévy exponent* of the subordinator and, of course, the Lévy exponent characterizes the law of the subordinator. The closure of the range of a subordinator is called a *regenerative set*. Moreover, it turns out to be the level set of a Markov process [Fitzsimmons *et al.* (1985)]. Note that $\{\sigma_{ct}\}_t$ has the same range, but apart for this degeneracy, different subordinators yield different processes with values in $\mathcal{C}_\infty$. And since if σ_t is replaced by σ_{ct} the Lévy exponent becomes $c\Phi(\cdot)$, we can (and will) set $\Phi(1) = 1$. A very important result proven in [Fitzsimmons *et al.* (1985)] is that if one has pointwise convergence of a sequence of Lévy exponents $\{\Phi_N(\cdot)\}_N$ to a limit that we call $\Phi(\cdot)$, then this limit is a Lévy exponent and one has convergence in law of the sequence of regenerative sets.

With this in hand it is rather easy to prove the following:

Theorem A.8 *If τ is a recurrent renewal process with inter-arrival discrete density $K(\cdot)$ and $\alpha \geq 0$ then the sequence of random sets $\{\tau/N\}_N$ converges as $N \to \infty$ to the regenerative set with Lévy exponent $\Phi(\lambda) = \lambda^{\min(\alpha,1)}$.*

Note in particular that if $\alpha \geq 1$ then $\sigma_t = t$ for every t and the limit regenerative set is simply $[0,\infty)$. Actually this part of the theorem does not need any kind of heavy machinery if $m_K < \infty$: τ is positive recurrent and the points in τ/N are typically at distance $O(1/N)$. For $\alpha = 0$ by regenerative set with Lévy exponent $\Phi(\lambda) = \lambda^0$ we simply mean the deterministic set $\{0\}$.

The limit process for $\alpha \in (0,1)$ is highly non-trivial and it has been studied in depth, see *e.g.* [Bertoin (1996)]. A result that we will need on its trajectories is that $\mathbf{P}(\max \mathcal{A}_\alpha \cap [0,1] = 1) = 0$, in fact $\max \mathcal{A}_\alpha \cap [0,1]$ is a continuous random variable supported on $[0,1]$ (whose density is explicitly known, see the *generalized arc-sine law* with parameter α, *e.g.* [Bingham *et al.* (1987), (8.6.0)]).

Proof. For $\alpha = 0$ we apply the arc-sine law for renewal theory, [Bingham *et al.* (1987), Theorem 8.6.5], which says (in particular!) that $\max \tau/N \cap [0,1] \Longrightarrow 0$, and of course $[0,1]$ may be substituted by $[0,t]$, any $t > 0$. For $\alpha > 0$ instead by direct computation we see that the Lévy exponent of τ/N is

$$\Phi_N(\lambda) = \gamma \sum_{n=1}^{\infty} (1 - \exp(-\lambda n/N))\, K(n). \tag{A.47}$$

Choose $\alpha = \alpha_N$ such that $\Phi_N(1) = 1$. It suffices to show the convergence of $\Phi_N(\lambda)$ to $\lambda^{\min(\alpha,1)}$. If $m_K < \infty$ this is an immediate consequence of the Dominated Converge Theorem. For $\alpha \in (0,1)$ we use the uniform converge property of slowly varying functions, (A.18), to see that for every $\varepsilon \in (0,1)$ as $N \to \infty$

$$\gamma_N \sum_{n=\lfloor N\varepsilon \rfloor}^{n=\lfloor N/\varepsilon \rfloor} (1 - \exp(-\lambda n/N))\, K(n) \sim \frac{\gamma_N}{N^{\alpha} L(N)} \int_{\varepsilon}^{1/\varepsilon} \frac{1 - \exp(-\lambda r)}{r^{1+\alpha}}\, dr, \tag{A.48}$$

and choose γ_N such that the pre-factor in the right-hand side converges to $c > 0$ (in fact, choose $c = \alpha/\Gamma(1-\alpha)$). By using property L.1 of slowly varying functions it is easy to see that small and large values of n missing in the sum in the left-hand side of (A.48) give no contribution when one takes the limit as $N \to \infty$ and then $\varepsilon \searrow 0$. Therefore

$$\lim_{N \to \infty} \Phi_N(\lambda) = c \int_0^{\infty} \frac{1 - \exp(-\lambda r)}{r^{1+\alpha}}\, dr = \lambda^{\alpha}. \tag{A.49}$$

We are therefore left with the case of $\alpha = 1$ and $m_K = \infty$. In this case it is easy to realize that the dominant part in the summation in (A.47) comes from small values of n, say $n < \varepsilon N$ for $\varepsilon > 0$. For these values of n we may replace $1 - \exp(-\lambda n/N)$ with $\lambda n/N$ and we are therefore left with

$$\lambda \frac{\gamma_N}{N} \sum_{n < \varepsilon N} \frac{L(n)}{n}. \tag{A.50}$$

By L.4, $\sum_{n<N} L(n)/n$ is slowly varying and therefore it behaves asymptotically $\sum_{n<\varepsilon N} L(n)/n$ as long as $\varepsilon > 0$. The obvious choice of γ_N completes the argument. $\qquad \square$

A.6   Random Walks

Here we present a number of facts on random walks. We start with the *reflection principle* for (p, q)-walks (see Section 1.2.1).

We have the following:

Proposition A.9   *For every $k \in \mathbb{N}$ and every $N \in \mathbb{N}$ we have*

$$\mathbf{P}\left(S_n \geq 0,\ n = 1, 2, \ldots, N - 1,\ S_N = k\right) =$$
$$\mathbf{P}\left(S_N = k\right) - \mathbf{P}\left(S_N = k + 2\right), \quad \text{(A.51)}$$

and

$$\mathbf{P}\left(S_n > 0,\ n = 1, 2, \ldots, N - 1,\ S_N = k\right) =$$
$$\frac{p}{2}\left[\mathbf{P}\left(S_{N-1} = k - 1\right) - \mathbf{P}\left(S_{N-1} = k + 1\right)\right]. \quad \text{(A.52)}$$

Proof. Note that the second statement follows from the first since in that case the first step of the walk is obliged, that is $S_1 = 1$, which happens with probability $p/2$ and then one is left with applying the first formula.

Let us then show (A.51). We have

$$\mathbf{P}\left(S_n \geq 0,\ n = 1, 2, \ldots, N - 1,\ S_N = k\right) =$$
$$\mathbf{P}\left(S_N = k\right) - \mathbf{P}\left(\min_{n=1,\ldots,N-1} S_n < 0,\ S_N = k\right) =$$
$$\mathbf{P}\left(S_N = k\right) - \mathbf{P}\left(S_N = -k - 2\right), \quad \text{(A.53)}$$

where the last step follows by conditioning on the stopping time $\inf\{n : S_n = -1\}$ and by using the strong Markov property and the symmetry of the walk. The symmetry is used again in the last line of (A.53) to conclude. $\square$

From Proposition A.9 one extracts a rather explicit expression for the law of $\tau := \inf\{n : S_n = 0\}$, in fact

$$\mathbf{P}\left(\tau > N\right) = 2\sum_{k > 0} \mathbf{P}\left(S_n > 0, n = 1, \ldots, N - 1,\ S_N = k\right), \quad \text{(A.54)}$$

so that, by (A.52), one obtains

$$\mathbf{P}\left(\tau > N\right) = p\left[\mathbf{P}\left(S_{N-1} = 0\right) + \mathbf{P}\left(S_{N-1} = 1\right)\right]. \quad \text{(A.55)}$$

Therefore the asymptotic behavior $\mathbf{P}\left(\tau > N\right) \overset{N\to\infty}{\sim} \sqrt{2p/\pi N}$ is an immediate consequence of the classical Local Limit Theorem (see Appendix A.1, formula (A.4): note that the variance of S_1 is p). Moreover for $N \in \mathbb{N}$ we have

$$\mathbf{P}\left(\tau = N + 1\right) = p \sum_{k=0,1} \left[\mathbf{P}\left(S_{N-1} = k\right) - \mathbf{P}\left(S_N = k\right)\right], \qquad (A.56)$$

and $\mathbf{P}\left(\tau = 1\right) = q$. If we now use the local limit expansion (A.5) we obtain that for every $k \in \mathbb{Z}$

$$\mathbf{P}\left(S_{N-1} = k\right) - \mathbf{P}\left(S_N = k\right) \overset{N\to\infty}{\sim} \frac{1}{2p}\sqrt{\frac{p}{2\pi N^3}}. \qquad (A.57)$$

We collect what we have just obtained in the following statement:

Proposition A.10 *For any (p,q)-walk, $p \in (0,1)$, we have*

$$\mathbf{P}\left(\tau = n\right) \overset{n\to\infty}{\sim} n^{-3/2}\sqrt{\frac{p}{2\pi}}. \qquad (A.58)$$

An alternative proof of Proposition A.10 may be found in [Caravenna *et al.* (2006a), Section A.3] where it is shown, based on a result of [Alili and Doney (1999)], that

$$\lim_{N\to\infty} N^{3/2}\mathbf{P}\left(S_n > 0,\, n = 1, 2, \ldots, N-1,\, S_N = 0\right) = c \in (0,\infty), \quad (A.59)$$

for general random walks with centered increments supported in $\mathbb{Z}$ (and not supported on any sub-lattice of $\mathbb{Z}$) and of finite non-degenerate variance (c is identified explicitly).

For completeness we mention that the result for the $(1,0)$-walk is instead $\mathbf{P}\left(\tau = 2n\right) \sim n^{-3/2}\sqrt{1/(4\pi)}$ [Feller (1966), Chapter III]).

Proposition A.10 is a particular case of a much stronger result, namely:

Theorem A.11 *[Kesten (1963), Theorem 8] If S is a random walk on $\mathbb{Z}$ such that*

- *for every $k \in \mathbb{Z}$ we have $\mathbf{P}(S_n = k) > 0$ for n sufficiently large;*
- *there exists $\beta \in [1,2]$ such that*

$$\lim_{t\to 0} \frac{1}{|t|^\beta}\left(1 - \mathbf{E}\left[\exp(itS_1)\right]\right) = Q \in (0,\infty); \qquad (A.60)$$

then

$$n^{2-(1/\beta)} \mathbf{P}(\tau = n) \overset{n \to \infty}{\sim} \frac{(\beta - 1)\sin(\pi/\beta)}{\Gamma(1/\beta)} Q^{1/\beta}, \qquad (A.61)$$

if $\beta \in (1, 2]$ and

$$n(\log n)^2 \mathbf{P}(\tau = n) \overset{n \to \infty}{\sim} \pi Q, \qquad (A.62)$$

if $\beta = 1$.

The proof of Theorem A.11 is based on the fact that the local limit expansion yields the asymptotic behavior of the Laplace transform of $\mathbf{P}(S_n = 0)$, which in turn, via formula (A.42), yields the asymptotic behavior of the Laplace transform of the distribution of τ. A Tauberian argument then gives directly the asymptotic behavior of $\mathbf{P}(\tau > n)$, but extracting the asymptotic behavior of $\mathbf{P}(\tau = n)$ is not obvious.

We have not been able to find the analog to Theorem A.11 for random walks in $d \geq 3$ in the literature, but Ron Doney [private communication (2006)] supplied us with such a result: if τ is the first return time to the origin of an aperiodic d–dimensional random walk with centered increments in L^2, then $\mathbf{P}(\tau = n) \overset{n \to \infty}{\sim} c/n^{d/2}$. Note that, as explained above, by the multi-dimensional Local Limit Theorem, (A.42) and by a Tauberian argument, one obtains (explicitly) the asymptotic behavior of $\mathbf{P}(\tau > n)$, which yields the constant $c(> 0)$ (that depends on d and on the covariance structure of the walk). We do not reproduce here Doney's proof, but we make it available upon request.

In $d = 2$ instead $\mathbf{P}(\tau = n) \overset{n \to \infty}{\sim} c/(n(\log n)^2)$ [Jain and Pruitt (1972)], with $c > 0$ once again explicit.

A.7   The Super-Additive Ergodic Theorem

A natural and very important probabilistic generalization of the well–known converge result for super-additive deterministic sequences is due to J. F. C. Kingman [Kingman (1973)]. Let us first recall the deterministic result: a sequence $\{a_n\}_n$ of real numbers is super-additive if $a_{n+m} \geq a_n + a_m$ for every n and $m \in \mathbb{N}$.

Proposition A.12   *If $\{a_n\}_n$ is a super-additive sequence, then the limit* *of $\{a_n/n\}_n$ exists, taking possibly the value $+\infty$, and*

$$\lim_{n\to\infty} \frac{a_n}{n} = \sup_{n\in\mathbb{N}} \frac{a_n}{n}. \tag{A.63}$$

It is worth recalling that, in spite of what (A.63) might suggest, in general neither $\{a_n\}_n$ nor $\{a_n/n\}_n$ are increasing. This is obvious, because linear sequences, choose *e.g.* $a_n = -n$, are super-additive. A less trivial example is given by $a_n = -\log(1+n)$.

Proof. Let us first consider the case $\sup_n a_n/n = +\infty$. In this case for every $c > 0$ one may find $n_0 \in \mathbb{N}$ such that $a_{n_0} \geq Cn_0$. Given n_0, any $n \in \mathbb{N}$ uniquely defines $k = k(n,n_0) \in \mathbb{N}\cup\{0\}$ and $m = m(n,n_0) \in \{0,\ldots,n_0-1\}$ such that $n = kn_0 + m$. Of course $\lim_{n\to\infty} n/k = n_0$. By super-additivity and the choice of n_0 we have

$$\frac{a_n}{n} \geq \frac{k}{n}a_{n_0} + \frac{a_m}{n} \geq \frac{k}{n}Cn_0 + \frac{a_m}{n}. \tag{A.64}$$

But $|a_m| \leq \max_{n=1,\ldots,n_0-1}|a_n| < \infty$, so that from (A.64) one immediately infers $\liminf_{n\to\infty} a_n/n \geq C$ and therefore, since C is arbitrary, $\lim_{n\to\infty} a_n/n = \infty$.

Consider now the case $s := \sup_n a_n/n < +\infty$: it suffices to show that $\liminf_{n\to\infty} a_n/n \geq s$. Note that for every $\varepsilon > 0$ one may find $n_0 \in \mathbb{N}$ such that $a_{n_0} \geq (s-\varepsilon)n_0$. By writing once again n as kn_0+m, the super-additive property immediately yields

$$\frac{a_n}{n} \geq \frac{k}{n}(s-\varepsilon)n_0 + \frac{a_m}{n}, \tag{A.65}$$

which entails $\liminf_{n\to\infty} a_n/n \geq s - \varepsilon$ and the proof is complete. $\square$

Proposition A.12 has a beautiful generalization, due to J. F. C. Kingman [Kingman (1973)], to random sequences.

We call *2–index process* the process $F := \{F_{i,j}\}_{i,j\in\mathbb{N}:\ i<j}$, so $F_{i,j}$ is a (measurable) function from Ω to $\mathbb{R}$.

Theorem A.13 *Let us assume that F is a super-additive process that is F is a 2–index process such that*

(1) for every choice of natural numbers k, l and $i_1 < j_1$, $i_2 < j_2,\ldots,$ $i_k < j_k$ we have

$$(F_{i_1,j_1}, F_{i_2,j_2},\ldots,F_{i_k,j_k}) \sim (F_{i_1+l,j_1+l}, F_{i_2+l,j_2+l},\ldots,F_{i_k+l,j_k+l}); \tag{A.66}$$

(2) for every $i < k < j$

$$F_{i,j} \geq F_{i,k} + F_{k,j}; \tag{A.67}$$

(3) $F_{1,j}$ is integrable for every j and the following bound holds

$$\sup_j \frac{1}{j}\mathbb{E}\left[F_{1,j}(\omega)\right] < \infty. \tag{A.68}$$

Then the limit as j tends to infinity of $F_{1,j}/j$ exists $\mathbb{P}(\,\mathrm{d}\omega)$–a.s. and in $L^1(\Omega, \mathcal{A}, \mathbb{P})$. The limit, that we denote by $f(\omega)$, is θ–invariant, in the sense that $f(\theta\cdot) = f(\cdot)$.

For a proof and generalizations we refer to [Kingman (1973)], [Krengel (1985)] and [Liggett (1985)].

A.8   Perron–Frobenius Theory

Let A be a $n \times n$ matrix with non-negative entries. A is said to be *irreducible* if for every $i, j \in \{1, \ldots, n\}$ there exists $m = m(i, j)$ such that $(A^m)_{i,j} > 0$ or, equivalently, if for every i and j there exist $k \in \mathbb{N}$ and indices $i =: i_1, i_2, \ldots, i_k := j$ such that $A_{i_{l-1}, i_l} > 0$ for $l = 2, \ldots, k$.

Theorem A.14   (Perron–Frobenius Theorem).   *If A is a non-negative irreducible matrix then A possesses a unique eigenvalue $\lambda_\star = \lambda_\star(A)$, called the Perron–Frobenius eigenvalue, such that*

(1) $\lambda_\star > 0$;
(2) if $\lambda \in \mathbb{C}$ is an eigenvalue of A, then $|\lambda| \leq \lambda_\star$;
(3) there exist left and right eigenvectors of A that are elements of $(0, \infty)^n$. These eigenvectors are unique up to multiplication by (positive) constants (and therefore $\lambda_\star$ is a simple eigenvalue).

A proof of Theorem A.14 may be found for example in [Minc (1988)] where one finds also the following formula for $\lambda_\star$:

$$\lambda_\star = \max_{\substack{v \in [0,1]^n \\ \sum_i v_i = 1}} \min_{j:\, v_j > 0} \frac{(Av)_j}{v_j}. \tag{A.69}$$

From this formula and the Perron–Frobenius Theorem one easily infers that if A is irreducible and B is a non-negative matrix with $B_{i,j} > 0$ for at least a choice of i and j, then $\lambda_\star(A + B) > \lambda_\star(A)$. Since $\lambda_\star$ is simple one easily obtains that if A is a smooth function of a real parameter (for example if $A_{i,j}$ is analytic for every i, j) then $\lambda_\star$ inherits the smoothness (in particular if the entries are analytic, $\lambda_\star$ is analytic too). If A' is the derivative of A and if u and v are the right and left eigenvalues associated to the Perron-Frobenius eigenvalue, with the normalization condition $u \cdot v = 1$, then $\lambda'_\star = u \cdot A'v$.

Another relevant result is that if the entries of A are log-convex functions of a parameter, then $\lambda_\star(A)$ is log-convex [Kingman (1961)]. We recall that a positive function g is log-convex if $\log g$ is convex (also $g(\cdot) \equiv 0$ is usually considered log-convex). Note that a log-convex function is also convex.

Appendix B

Some Technical Estimates

B.1 Homogeneous Localization Strategy: The Entropy Density

For $b \geq 0$ we set $K_b(n) := \exp(-bn)K(n)/\Theta(b)$ and $\Theta(b) := \sum_n \exp(-bn)K(n)$. With reference to Definition 1.4, we assume $\Sigma_K = 1$, so $K(\cdot)$ is really a probability. Note that $\Theta(\cdot) \in C^\infty$ on the positive semi-axis and that $-(\log \Theta(b))'$ is equal to m_{K_b}. We consider the law of the renewal processes associated to $K(\cdot)$ and to $K_b(\cdot)$ up to time N, that is the law of the random set $\tau \cap \{0, 1, \dots, N\}$ (below we will write simply τ). Of course these are probability measures on the finite set

$$\cup_{k=0}^N \left\{ \underline{n} \in \mathbb{N}^{k+1} : 0 = n_0 < n_1 < \dots n_k \leq N \right\}. \tag{B.1}$$

We call $\nu_{b,N}$ the law of $\tau \cap \{0, 1, \dots, N\}$ when the inter-arrival times are distributed like $K_b(\cdot)$. We want to compute the relative entropy density of $\nu_{b,N}$ with respect to $\nu_{0,N}$.

Proposition B.1 *With the notation $\mathcal{S}_{N,b} := \mathcal{S}(\nu_{b,N}|\nu_{0,N})$ we have that for every $b > 0$*

$$\lim_{N \to \infty} \frac{1}{N} \mathcal{S}_{N,b} = -b + \Theta(b)\frac{\log \Theta(b)}{\Theta'(b)} =: \mathrm{s}(b). \tag{B.2}$$

A look at the proof suffices to realize that the precise requirements on the asymptotic decay of $K(\cdot)$ given in Definition 1.4 are not needed and they may be replaced by the requirement that $K(\cdot)$ tends to zero slower than any exponential, namely that $K(n) \exp(\varepsilon n) \overset{n \to \infty}{\longrightarrow} \infty$ for every $\varepsilon > 0$.

Proof. Fix $b > 0$. With $k := \mathcal{N}_N(\tau)$, recall (A.30), we have

$$\frac{d\nu_{b,N}}{d\nu_{0,N}}(\tau) = \left(\prod_{i=1}^{k} \frac{\exp\left(-b(\tau_i - \tau_{i-1})\right)}{\Theta(b)} \right) R_{N,b}(\tau), \qquad (B.3)$$

where the term between parentheses in the right-hand side takes the value 1 if $k = 0$ and

$$R_{N,b}(\tau) := \frac{\sum_{n>N-\tau_k} K_b(n)}{\sum_{n>N-\tau_k} K(n)}. \qquad (B.4)$$

Denote by $\mathbf{P}_b$ the law of the renewal process with inter-arrival distribution $K_b(\cdot)$. We will show below that

$$\lim_{N\to\infty} \frac{1}{N} \mathbf{E}_b \left[\log R_{N,b}(\tau) \right] = 0. \qquad (B.5)$$

Let us focus on

$$\frac{1}{N} \mathbf{E}_b \left(\log \prod_{i=1}^{\mathcal{N}_N(\tau)} \frac{\exp\left(-b(\tau_i - \tau_{i-1})\right)}{\Theta(b)} \right) = -b\frac{\tau_{\mathcal{N}_N(\tau)}}{N} - \frac{\mathcal{N}_N(\tau)}{N} \log \Theta(b). \qquad (B.6)$$

By (A.31) and the Lebesgue Dominated Convergence Theorem the second term in the right-hand side converges as $N \to \infty$ to $\log \Theta(b)/m_{K_b}$, which is equal to $-\Theta(b) \log \Theta(b)/\Theta'(b)$. For the second term we observe that $\tau_{\mathcal{N}_N(\tau)} = N - A_N(\tau)$, where $A(\tau)$ is the backward recurrence time process associated to τ, see Appendix A.5. But, as pointed out in (A.29), the sequence $A(\tau)$ converges in law and, since $0 \le A_N(\tau)/N \le 1$, again by the Dominated Convergence Theorem we have that $\mathbf{E}_b \left[A_N(\tau) \right]/N = 0$. Therefore the first term in the right-hand side of (B.6) converges to $-b$.

We are therefore left with proving (B.5). To this purpose first observe that

$$R_{N,b}(\tau) \le \left(\sum_{n>N-\tau_{\mathcal{N}_N(\tau)}} K(n) \right)^{-1} \le \left(\sum_{n>N} K(n) \right)^{-1} \le 1/K(N), \quad (B.7)$$

and, since $\log K(N)/N \to 0$, we have that $\limsup_{N\to\infty} \mathbf{E}[\log R_{N,b}(\tau)/N] \le 0$.

For the lower bound we observe instead that

$$R_{N,b}(\tau) \ge \sum_{n>N-\tau_{\mathcal{N}_N(\tau)}} K_b(n) \ge c \exp\left(-2b(N - \tau_{\mathcal{N}_N(\tau)})\right), \qquad (B.8)$$

for some positive constant c depending on b and $K(\cdot)$. The $\mathbf{P}_b$–expectation of the logarithm of the rightmost side of (B.8) vanishes as $N \to \infty$ by the same argument used for the first term in the right-hand side of (B.6), so that $\liminf_{N\to\infty} \mathbf{E}[\log R_{N,b}(\tau)/N] \geq 0$ and the proof is complete. $\square$

B.2 The Weak Localization Limit

We are now going to concentrate ourselves on the asymptotic properties of $s(b)$, as $b \searrow 0$. We assume that $K(\cdot)$ is chosen as in Definition 1.4 and $\Sigma_K = 1$. The main result is the following:

Proposition B.2 *For $\alpha > 0$ and $b \searrow 0$ we have*

$$s(b) \sim \frac{\max((1 - \alpha), 0)}{\alpha} b. \tag{B.9}$$

Note that, for $\alpha \geq 1$, (B.9) simply says that $s(b) = o(b)$. The case $\alpha = 0$ is treated below.

Proof. The result for $\alpha \geq 1$ and $m_K < \infty$ is a direct consequence of $\Theta(b) \sim 1$ and $\Theta'(b) \sim -m_K$. Let us therefore assume that $\alpha \in (0, 1]$ and $m_K = \infty$. We are going to use summation by parts and we first observe that

$$\overline{K}(n) := \sum_{j>n} K(j) \overset{n\to\infty}{\sim} \frac{L(n)}{\alpha\, n^\alpha}, \tag{B.10}$$

where the definition is given for every $n \in \mathbb{N} \cup \{0\}$ and for the asymptotic behavior we have applied (A.22). Note that $\overline{K}(0) = 1$. The following identity holds:

$$\Theta(b) = 1 - (1 - \exp(-b)) \sum_{n=0}^{\infty} \exp(-bn)\overline{K}(n), \tag{B.11}$$

so that

$$\frac{1 - \Theta(b)}{b} \sim \sum_{n=0}^{\infty} \exp(-bn)\overline{K}(n)$$

$$\sim \frac{1}{\alpha} \sum_{n=1}^{\infty} n \exp(-bn)K(n) = -\frac{1}{\alpha}\Theta'(b), \tag{B.12}$$

where in the second asymptotic estimate we have used (B.10) and $m_{K_b} < \infty$. Since of course $\Theta(b) \sim 1$, we obtain

$$\Theta(b) \frac{\log \Theta(b)}{\Theta'(b)} \sim \frac{1}{\alpha} b, \tag{B.13}$$

and the proof is complete.   □

Note that in the previous argument we have avoided using the precise asymptotic behavior of $\Theta'(b)$ when $m_K = \infty$ (we have actually used $\Theta'(b) \overset{b \searrow 0}{\longrightarrow} \infty$). However combining (A.21) and Theorem A.2 one gets

$$m_{K_b} = -\Theta'(b) \overset{b \searrow 0}{\sim} \frac{\Gamma(2 - \alpha)}{1 - \alpha} b^{\alpha - 1} L(1/b), \tag{B.14}$$

for $\alpha \in [0, 1)$ (and recall that $\Gamma(2 - \alpha)/(1 - \alpha) = \Gamma(1 - \alpha)$).

Let us discuss the marginal cases ($\alpha = 0$ or 1):

- For $\alpha = 1$ and $m_K = \infty$ one can show directly that the map $x \mapsto m_{K_{1/x}}$ is slowly varying [Bingham *et al.* (1987), p. 10–11].
- For $\alpha = 0$ the asymptotic behavior $m_{K_b} \sim L(1/b)/b$ is given by (B.14). Notice however that in Proposition B.2 the case $\alpha = 0$ has been left out, but (B.9) suggests that $s(b) \gg b$. Given the formula (B.2) for $s(b)$ it is clear that it suffices to look at the term $\Theta(b) \log \Theta(b)/\Theta'(b)$, which coincides with $-\log \Theta(b)/m_{K_b}$. Note that by L.4 of Appendix A.4 and by Theorem A.2 one has

$$1 - \Theta(b) \overset{b \searrow 0}{\sim} \widetilde{L}(1/b), \tag{B.15}$$

where $\widetilde{L}(\cdot)$ is a slowly varying function, vanishing at infinity, such that $\lim_{x \to \infty} \widetilde{L}(x)/L(x) = \infty$. Therefore we have $s(b) \sim b \widetilde{L}(1/b)/L(1/b)$.

B.3   Localization and Loss of Memory

This section contains the proof of Lemma 7.10 and from this lemma we take the framework and the notations. We start off by giving four lemmas (the first two are used to prove the third and the last two are used in the main proof). Then we give the proof.

Lemma B.3   *For every $(\beta, h) \in \mathcal{L}$ and for every $\varepsilon > 0$ there exists $c > 0$ such that for every N, every u and l in $\mathbb{N}$, with $0 < u < l < N$, and every*

$B \subset \{u, \ldots, l\}$, with $u, l \in B$, we have

$$\mathbb{E}\mathbf{P}^c_{N,\omega}(\tau \cap [u, l] = B) \leq c^{|B|-1} \exp\left(-(\mu(\beta, h) - \varepsilon)(l - u)\right). \quad (B.16)$$

Proof. Let us write B as $\{n_1, \ldots, n_k\}$, with $u = n_1 < n_2 < \ldots < n_k = l$. We have

$$\mathbf{P}^c_{N,\omega}(\tau \cap [u, l] = B) =$$

$$\frac{1}{Z^c_{N,\omega}} Z^c_{n_1,\omega} \left(\prod_{j=2}^{k} K(n_j - n_{j-1}) \exp\left(\beta\omega_{n_j} - h\right)\right) Z^c_{N-n_k, \theta^{n_k}\omega}, \quad (B.17)$$

and by replacing the denominator with $Z^c_{N,\omega}(B \subset \tau)$, which can be factorized in a way similar to what has been done in the numerator of the right-hand side of (B.17), we obtain the bound

$$\mathbf{P}^c_{N,\omega}(\tau \cap [u, l] = B) \leq \prod_{j=2}^{k} \frac{K(n_j - n_{j-1}) \exp\left(\beta\omega_{n_j} - h\right)}{Z^c_{n_j - n_{j-1}, \theta^{n_{j-1}}\omega}}. \quad (B.18)$$

By Theorem 7.3, for every $\varepsilon > 0$ we can find $c > 0$ such that for every n

$$\mathbb{E}\left[\frac{K(n) \exp\left(\beta\omega_n - h\right)}{Z^c_{n,\omega}}\right] \leq c \exp\left(-(\mu(\beta, h) - \varepsilon)n\right). \quad (B.19)$$

At this point it suffices to observe that the factors in the right-hand side of (B.18) are independent and by (B.19) we conclude. $\quad\square$

Lemma B.4 *Let $m \in \mathbb{N}$ and $i_0, \ldots, i_m$ such that $0 \leq i_0 < i_1 < \ldots < i_m \leq N$. Let moreover E_j, $j = 0, \ldots, m$, be an event in the σ-algebra generated by $\tau \cap (i_j, i_{j+1}]$. Then*

$$\mathbf{P}^c_{N,\omega}\left(\{i_0, \ldots, i_m\} \in \tau, \cap_{j=0}^{m-1} E_j\right) \leq \prod_{j=0}^{m-1} \mathbf{P}^c_{i_{j+1} - i_j, \theta^{i_j}\omega}(E_j). \quad (B.20)$$

Proof. This follows directly by writing the left-hand side in (B.20) as ratio of partition functions, the one in the numerator is restricted to the event we are considering. The denominator is bounded from below by inserting the event $\{i_0, \ldots, i_m\} \in \tau$. By the renewal property we conclude. $\quad\square$

In what follows $c := \mu(\beta, h)/2$ and we call T_0 the renewal point in τ closest to u, but smaller than u, that is $\max\{\tau \cap [0, u)$. $T_1 < T_2 < \ldots$ are the renewals to the right of T_0. Set $r := \sup\{j \geq 0 : T_j \leq l\}$, so T_r is the right-most renewal smaller or equal to l. Set also $\chi_j := T_{j+1} - T_j$: we call j^{th} excursion the interval $T_j, T_j + 1, \ldots, T_{j+1}$, so that χ_j is the length of the jth excursion. Note that the 0th and r^{th} excursions have one endpoint outside of $\{u, u + 1, \ldots, l\}$.

We have the following:

Lemma B.5   *There exist two positive constants κ_1 and κ_2 such that for every $(\beta, h) \in \mathcal{L}$ there exists $C_1 := C_1(\beta, h)$ such that*

$$\mathbf{E}\mathbf{P}_{N,\omega}^{\mathrm{c}}\left(\sum_{\substack{j:\, 0 \leq j \leq r \\ \chi_j \geq \kappa_2/\mathrm{C}}} \chi_j \geq (l - u)/4 \right) \leq C_1 \exp\left(-(l - u)\kappa_1 \mathrm{C}\right), \qquad \text{(B.21)}$$

for every N and every u and l, with $0 \leq u < l \leq N$.

Proof. We start by setting

$$\ell := \sum_{\substack{j:\, 0 < j < r \\ \chi_j \geq \kappa_2/\mathrm{C}}} \chi_j, \qquad \text{(B.22)}$$

and $x := \chi_1$, as well as $y := \chi_r$, so that the event of which we have to estimate the probability is characterized by the condition

$$\ell + x\mathbf{1}_{x \geq \kappa_2/\mathrm{C}} + y\mathbf{1}_{y \geq \kappa_2/\mathrm{C}} \geq (l - u)/4. \qquad \text{(B.23)}$$

When the positive integer numbers x, y and ℓ satisfy (B.23) and $\ell \leq l - u$ we simply write $(x, y, \ell) \in D(u, l)$. If τ is in the event of which we have to evaluate the probability, the number, that we call n, of long excursions (*i.e.* excursions that are long at least κ_2/C) contained in $\{u, \ldots, l\}$ is at most $\lfloor \ell \mathrm{C}/\kappa_2 \rfloor$. The number of ways that such excursions can be put in $\{u, \ldots, l\}$ is bounded above (in a very rough way) by the number of ways in which one can choose the left and right endpoints of the ℓ excursions, dropping the order constraints, obtaining thus

$$\binom{l - u}{n}^2. \qquad \text{(B.24)}$$

At this point we apply Lemma B.3 and Lemma B.4 to obtain that the left-hand side of (B.21) is bounded above by

$$\sum_{(x,y,\ell)\in D(u,l)} (l-u)^2 \exp\big(-\mathrm{c}(\ell+x+y)\big) \sum_{n=0}^{\lfloor \ell \mathrm{c}/\kappa_2 \rfloor} \binom{l-u}{n}^2 c_1^{n+1}, \quad \text{(B.25)}$$

where c_1 is the constant appearing in Lemma B.3 and the factor $(l-u)^2$ takes into account the number of ways one can place the right (respectively left) endpoint of the 0th (respectively r^{th}) excursion. Since $\ell \leq l-u$, by Stirling formula we have that for every $\varepsilon > 0$ we can choose κ_2 (sufficiently large) that the sum over n is bounded above by $((l-u)\mathrm{c}/\kappa_2)\exp(-\varepsilon(l-u))$ for every $\ell(\leq l-u)$. One then performs the sum over $(x,y,\ell) \in D(u,l)$, for which $\ell + x + y \geq (l-u)/4$, and from this one concludes. $\qquad\square$

Lemma B.6 *For every n there exists $\varepsilon_n > 0$ such that for every $I \subset \{1,\dots,n\}$ and every ω we have*

$$\frac{\mathbf{P}^c_{n,\omega}(\tau \cap \{1,\dots,n-1\} = \emptyset)}{\mathbf{P}^c_{n,\omega}(\tau \cap (\{1,\dots,n-1\}\backslash I) = \emptyset)} \leq \prod_{i\in I} \frac{1}{1 + \varepsilon_n \exp(\beta\omega_i - h)}. \quad \text{(B.26)}$$

Proof. With the notation $\widetilde{\zeta}(a) = \exp(\beta a - h)$ we can write the left-hand side in (B.26) as

$$\frac{K(n)\widetilde{\zeta}(\omega_n)}{\sum_{A\subset I} \prod_{i=1}^{|A|+1} \widetilde{\zeta}(\omega_{a_i}) K(a_i - a_{i-1})}, \quad \text{(B.27)}$$

where, given A, $a_0 := 0$, $a_{|A|+1} = n$, $A = \{a_1,\dots,a_{|A|}\}$, with $a_j > a_{j-1} > 0$ for every j. In the sum over $A \subset I$ is included $A = \emptyset$. We set

$$\varepsilon_n := \min_{A\subset\{1,\dots,n-1\}} \left(\frac{\prod_{i=1}^{|A|+1} K(a_i - a_{i-1})}{K(n)}\right)^{1/(|A|+1)}, \quad \text{(B.28)}$$

and of course $\varepsilon_n > 0$. Therefore the term in the left-hand side of (B.26) is bounded above by

$$\left(\sum_{A\subset I} \prod_{i=1}^{|A|} \varepsilon_n \widetilde{\zeta}(\omega_{a_i})\right)^{-1}, \quad \text{(B.29)}$$

where if $A = \emptyset$ then the product over i is equal to 1. Performing the summation over A leads to (B.26). $\qquad\square$

We are now ready to prove Lemma 7.10. The proof goes as follows: by Lemma B.5 with large probability a substantial portion, at least three quarters, of $\{u, \ldots, l\}$ is covered by short excursion (*i.e.* shorter than κ_2/C) and this is of course true for both independent copies, $\tau^{(1)}$ and $\tau^{(2)}$ of the process. By large probability here we mean that the complementary event has a probability that is exponentially small in $l - u$. Therefore the intersection of these two random sets covers at least half of $\{u, \ldots, l\}$ and there will be several, *i.e.* a number proportional to $l - u$, $\tau^{(1)}$–contacts in short $\tau^{(2)}$–excursions (or the same is true exchanging 1 with 2). We then make a rough estimate to show that is improbable, with an exponential estimate, that this happens at the same time as $\tau^{(1)} \cap \tau^{(2)} \cap [u, l] = \emptyset$.

Proof of Lemma 7.10. We set

$$B_{u,l} := \left\{ (\tau^{(1)}, \tau^{(2)}) : \tau^{(1)} \cap \tau^{(2)} = \emptyset \right\}, \qquad (B.30)$$

and

$$E^{(s)} := \left\{ (\tau^{(1)}, \tau^{(2)}) : \sum_{\substack{j : 0 \le j \le r^{(s)} \\ \chi_j^{(s)} < \kappa_2/\mathrm{C}}} \chi_j^{(s)} > \frac{3}{4}(l - u) \right\}, \qquad (B.31)$$

with $s = 1, 2$ and obvious definition of $\chi^{(s)}$ and $r^{(s)}$. If we now define $\overline{E} := E^{(1)} \cap E^{(2)} \cap B_{u,l}$, by Lemma B.5 we have

$$\mathbb{E}\mathbf{P}_{N,\omega}^{\otimes 2}(B_{u,l}) \le 2C_1 \exp\left(-\kappa_1 \mathrm{C}(l - u)\right) + \mathbb{E}\mathbf{P}_{N,\omega}^{\otimes 2}\left(\overline{E}\right). \qquad (B.32)$$

We are therefore left with estimating the last term in the right-hand side. To do this we introduce the set $\mathbf{S}^i$, which is the union of the short excursions of $\tau^{(i)}$, and we observe that for $(\tau^{(1)}, \tau^{(2)}) \in \overline{E}$ the set $\mathbf{S}^1 \cap \mathbf{S}^2 \cap \{u, \ldots, l\}$ contains at least $(l - u)/4$ sites. Taking into account the upper bound on the length of a short excursion, the set

$$V := \left(\mathbf{S}^1 \cap \mathbf{S}^2 \cap \{u, \ldots, l\}\right) \cap \left(\tau^{(1)} \cup \tau^{(2)}\right), \qquad (B.33)$$

contains at least $\lfloor \mathrm{C}(l - u)/8\kappa_2 \rfloor$ points: note in particular that if $j \in V$ then either $j \in \tau^{(1)}$ and j belongs to an excursion of $\tau^{(2)}$, or the same holds exchanging $\tau^{(1)}$ and $\tau^{(2)}$. V is naturally written as disjoint union of W^1 and W^2, with $W^i := V \cap \tau^{(i)}$ and $\max_{i=1,2} |W^i| \ge \lfloor \mathrm{C}(l - u)/16\kappa_2 \rfloor$ which, by symmetry, implies

$$\mathbf{P}_{N,\omega}^{\otimes 2}\left(\overline{E}\right) \leq 2\mathbf{P}_{N,\omega}^{\otimes 2}\left(\overline{E} \cap \left\{|W^1| \geq c(l-u)/16\kappa_2\right\}\right). \qquad (B.34)$$

Now the situation is the following: by definition, a point in W^1 is in a short excursion of $\tau^{(2)}$. So there is of course no other $\tau^{(2)}$ point in the excursion, but the the intersection between this excursion and $\tau^{(1)}$ may contain more than one point. We therefore introduce the set $\hat{W}^1 \subset \{u,\ldots,l\} \cap \tau^{(1)}$ with the property that for every $j \in \hat{W}^1$ we can find x and y in $\tau^{(2)}$, with the properties that

(1) $u \leq x < j < y \leq l$ and $y - x \leq \kappa_2/c$;
(2) $\{x+1,\ldots,y-1\} \cap \tau^{(2)}$ is a subset of $\tau^{(1)}$.

These two properties do not imply that $\{x, x+1, \ldots, y\}$ is a short excursion of $\tau^{(2)}$ (this is due to the particular formulation of property (2) which leaves the freedom to $\tau^{(2)}$ to *touch down* on $\tau^{(1)}$) but on $\overline{E}$ they do, so that

$$\overline{E} \cap \left\{|W^1| \geq c(l-u)/16\kappa_2\right\} \subset \overline{E} \cap \left\{|\hat{W}^1| \geq c(l-u)/16\kappa_2\right\}, \qquad (B.35)$$

and therefore we can replace W^1 with $\hat{W}^1$ in (B.34). By conditioning we write

$$\mathbf{P}_{N,\omega}^{\otimes 2}\left(\overline{E}\right) \leq 2\mathbf{E}_{N,\omega}^{\otimes 2}\left[\mathbf{1}_{E^{(1)}}\mathbf{P}_{N,\omega}^{\otimes 2}\left(E^{(2)} \cap B_{u,l}\big|\tau^{(1)}\right)\right], \qquad (B.36)$$

and now we relax the condition $(\tau^{(1)}, \tau^{(2)}) \in B_{u,l}$ evaluating explicitly, by means of Lemma B.6, the probability change. We get

$$\mathbf{P}_{N,\omega}^{\otimes 2}\left(\overline{E}\right)$$
$$\leq 2\mathbf{E}_{N,\omega}^{\otimes 2}\left[\mathbf{1}_{E^{(1)}}\mathbf{E}_{N,\omega}^{\otimes 2}\left(\mathbf{1}_{E^{(2)}}\prod_{j \in \hat{W}^1}\left(\frac{1}{1 + \varepsilon\exp(\beta\omega_j - h)}\right)\Big|\tau^{(1)}\right)\right], \qquad (B.37)$$

where $\varepsilon := \varepsilon_{\lfloor\kappa_2/c\rfloor}$. If $\inf_j \varepsilon\exp(\beta\omega_j - h) > 0$ then the proof is complete, since $|\hat{W}^1| \geq \lfloor c(l-u)/16\kappa_2\rfloor$. This is true when ω_j is bounded from below. In the general case instead we observe that for κ_3 sufficiently large there exist positive constants c_1 and c_2 such that

$$\mathbb{P}\left(|\{i \in \{u,\ldots,l\} : \omega_i < -\kappa_3\}| > \frac{c(l-u)}{32\kappa_2}\right) \leq c_2\exp(-c_3(l-u)), \qquad (B.38)$$

for every $l - u$. The bound (B.38) is a standard Large Deviations estimate on the binomial variable with parameters $p := \mathbb{P}(\omega_1 < -\kappa_3)$ and $n = l - u$ (it holds for example if $p \leq C/64\kappa_2$). Therefore in estimating $\mathbb{E}\mathbf{P}_{N,\omega}^{\otimes 2}\left(\overline{E}\right)$ it suffices to consider disorder configurations ω such that $\omega_i \geq -\kappa_3$ outside of (at most) $C(l - u)/32\kappa_2$ sites. This guarantees that at least $\lfloor C(l - u)/32\kappa_2 \rfloor$ sites j in each (allowed) configuration of the set $\hat{W}^1$ are such that $\exp(\beta\omega_j - h) \geq \exp(-\beta\kappa_3 - h)$, and the proof is complete.

Lemma 7.10
□

Appendix C

Effective Interface Models

C.1 Ising Model, Effective Interface Models and Random Walks

We briefly review the link between interfaces in two-dimensional *bulk* systems and random walks, and the corresponding pinning and wetting problems. We just work out a precise example in order to give a glimpse of the extremely rich world of interface phenomena and models. For very complete presentations (also on higher dimensional systems) we refer to the classical works [Abraham (1986)] and [Fisher (1984)] and to the recent survey paper [Velenik (2006)].

An interface in a physical system is a surface that separates two materials or two phases of the same material. In general one does not pass from one phase to the other in an abrupt way, so one has to be more precise in defining what an interface is, but in some model systems, notably in the Ising system that we consider here, this can be done in an elementary way.

We consider an *anisotropic Ising model* on a box $\Lambda = \{1, \ldots, L_1\} \times \{1, \ldots, L_2\}$, that is the probability measure $\mathbf{P}_\Lambda^{J_1, J_2}$ concentrated on the configurations $\sigma \in \{-1, +1\}^{\Lambda \cup \partial \Lambda}$ such that $\sigma_{(x_1, 0)} = +1$ for $x_1 = 0, \ldots, L_1 + 1$ and $\sigma_{(x_1, x_2)} = -1$ for any other (x_1, x_2) in the external boundary $\partial \Lambda$ of Λ, and defined by

$$\mathbf{P}_\Lambda^{J_1, J_2}(\sigma) \propto \exp\left(-H_\Lambda^{J_1, J_2}(\sigma)\right), \tag{C.1}$$

where

$$H_\Lambda^{J_1, J_2}(\sigma) := -\frac{1}{2} \sum_{i=1,2} J_i \sum_{(x,y) \in A_i} \sigma_x \sigma_y, \tag{C.2}$$

and A_i is the set of couples $(x, y) \in \mathbb{Z}^2$ such that $\{x, y\} \cap \Lambda \neq \emptyset$, $|x - y| = 1$

and such that $|x_i - y_i| = 1$. J_1 and J_2 are positive parameters.

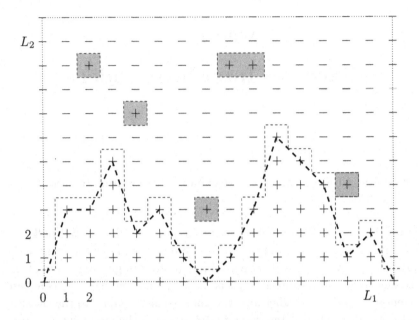

Fig. C.1 A sketch of the procedure leading from a two-dimensional bulk model, that is a model containing the details of the phases, to a $(1+1)$-dimensional effective interface model. An Ising spin configuration is represented (the dot line marks the boundary of the system). In the limit of $J_2 \to \infty$ this configuration has zero probability, but if we flip the spins in the six shadowed squares we obtain a positive probability configuration. And in this latter configuration the interface is reduced to a random walk bridge (thick dashed line).

In words, two spins interact if they are nearest neighbors and they interact differently in the vertical and horizontal directions. Note that in the definition of $H_\Lambda^{J_1,J_2}(\sigma)$ one can replace $\sigma_x \sigma_y$ with $-(\sigma_x - \sigma_y)^2/2$, up to a constant that does not modify the definition of the measure, so the effective energetic contributions are present only when there is a ± 1 contact, that is at the thin dashed line in Figure C.1. More precisely, up to a constant, $H_\Lambda^{J_1,J_2}(\sigma)$ is equal to $2J_1$ times the total length of the vertical segments of the thin dashed line plus $2J_2$ times the total length of vertical segments of the same line. We call interface the thin dashed line and we see that it has more than one connected component and only one component is an open line.

It is now clear that as J_2 tends to infinity the configuration in the figure has vanishing probability. A configuration with positive probability is the one in which the +'s that are put in evidence by a shadowed square are transformed into −'s. This is because as $J_2 \to \infty$ the only *surviving* configurations are the ones with minimal horizontal length of the dashed line (this minimal length is $L_1 + 1$) and the interface is reduced to one component. In this limit the interface is effectively a function, or a random walk configuration (the thick dashed line in Figure C.1).

Since the residual energy, after having taken the limit, is just given by the vertical length of the dashed line and since the configurations have the constraint to be non-negative and not to go beyond height L_2, the thick dashed line is a random walk, pinned at the boundary, which performs jumps of integer size t with probability proportional to $\exp(-2J_1|t|)$. Therefore, in the limit $L_2 \nearrow \infty$, the interface is precisely described by the wetting model of Section 1.3, at $\beta = 0$, that is the case in which only the hard wall constraint is present.

The example that we have worked out can be made richer in a number of ways:

(1) By adding a boundary field to the Ising Hamiltonian, that is a term of the type $\sum_{x_1=1}^{L_1} h_{x_1} \sigma_{(x_1,1)}$ with $h_1, \ldots, h_{L_1}$ real numbers, one obtains in the effective interface limit a (inhomogeneous) pinning energy and, in the homogeneous case, we are facing exactly the general wetting model of Section 1.3 and Figure 1.7.
(2) By adding a bulk field $h \sum_{x \in \Lambda} \sigma_x$ one favors or penalizes the − phase, according to the sign of h. One easily sees that this reflects on the effective model in terms of a linear potential, a *force*, acting on the interface (see [Hryniv and Velenik (2004)] and references therein).
(3) More general boundary conditions are naturally introduced and, for instance, one obtains in a straightforward ways models without hard walk constraint.

Bibliography

Abraham, D. B. (1986). Surface Structures and Phase Transitions, Exact Results, in *Phase Transitions and Critical Phenomena* **10**, Academic Press, London (UK), pp. 1–74.

Aizenman, M. and Wehr, J. (1990). Rounding Effects of Quenched Randomness on First-order Phase Transitions, *Commun. Math. Phys.* **130**, pp. 489–528.

Albeverio, S. and Zhou, X. Y. (1996). Free Energy and Some Sample Path Properties of a Random Walk with Random Potential, *J. Statist. Phys.* **83**, pp. 573–622.

Alexander, K. S. and Sidoravicius, V. (2006). Pinning of Polymers and Interfaces by Random Potentials, *Ann. Appl. Probab.* **16**, pp. 636–669.

Alili, L. and Doney, R. A. (1999). Wiener-Hopf Factorization Revisited and Some Applications, *Stoc. Stoc. Rep.* **66**, pp. 87–102.

Ané, C., Blachère, S., Chafaï, D., Fougères, P. , Gentil, I.,Malrieu, F., Roberto, C. and Scheffer, G. (2000). Sur les Inégalités de Sobolev Logarithmiques, Panoramas et Synthèses **10**, Société Mathématique de France, Paris.

Asmussen, S. (2003). Applied Probability and Queues, Second Edition, Applications of Mathematics **51**, Springer-Verlag, New York.

Bertoin, J. (1996). Lévy Processes, Cambridge Tracts in Mathematics, **121**. Cambridge University Press, Cambridge.

Bertoin, J. (1999). Subordinators: Examples and Applications, Lectured on probability theory and statistics (Saint-Flour, 1997), pp. 1–91, *Lecture Notes in Mathematics* **1717**, Springer-Verlag, Berlin (D).

Beylkin, G. and Monzón, L. (2005). On Approximation of Functions by Exponential Sums, *Appl. Comput. Harmon. Anal.* **19**, pp. 17–48.

Berenhaut, K. S. and Lund, R. B. (2002). Renewal Convergence Rates for DHR and NWU Lifetimes, *Probability in the Engineering and Informational Sciences* **16**, pp. 67–84.

Billingsley, P. (1968). Convergence of Probability Measures, John Wiley & Sons Inc., New York.

Bingham, N. H., Goldie, C. M. and Teugels, J. L. (1987). Regular Variation, Cambridge University Press, Cambridge.

Biskup, M. and den Hollander, F. (1999). A Heteropolymer near a Linear Interface, *Ann. Appl. Probab.* **9**, pp. 668–687.

Blossey, R. and Carlon, E. (2003). Reparametrizing the Loop Entropy Weights: Effect on DNA Melting Curves, *Phys. Rev. E* **68**, 061911 (8 pages).

Bodineau, T. and Giacomin, G. (2004). On the Localization Transition of Random Copolymers near Selective Interfaces, *J. Statist. Phys.* **117**, pp. 801–818.

Bolthausen, E. (2002). Large Deviations and Interacting Random Walks, Lectures on probability theory and statistics (Saint-Flour, 1999), pp. 1–124, *Lecture Notes in Math.* **1741**, Springer-Verlag, Berlin (D).

Bolthausen, E. and Giacomin, G. (2005). Periodic Copolymers at Selective Interfaces: a Large Deviations Approach *Ann. Appl. Probab.* **15**, pp. 963–983.

Bolthausen, E. and den Hollander, F. (1997). Localization Transition for a Polymer near an Interface, *Ann. Probab.* **25**, pp. 1334–1366.

Bolthausen, E. and Sznitman, A.-S. (2002). Ten Lectures on Random Media, *DMV Seminar* **32**, Birkhäuser Verlag, Basel (CH).

Bovier, A. (2006). Statistical Mechanics of Disordered Systems: a Mathematical Perspective, *Cambridge Series in Statistical and Probabilistic Mathematics* **18**, Cambridge University Press (UK).

Bovier, A. and Külske, C. (1996). There Are No Nice Interfaces in $(2 + 1)$-dimensional SOS Models in Random Media, *J. Statist. Phys.* **83**, pp. 751–759.

Bryn-Jones, A. and Doney, R. A. (2004). A Functional Central Limit Theorem for Random Walks Conditional to Stay Non-negative, *Proc. London Math. Soc.* (to appear).

Bundschuh, R. and Hwa, T. (2002). Statistical Mechanics of Secondary Structures Formed by Random RNA Sequences, *Phys. Rev. E* **65**, 031903.

Burkhardt, T.W. (1981). Localization-Delocalization Transition in a Solid-on-solid Model with a Pinning Potential, *J. Phys. A: Math. Gen.* **14**, pp. L63–L68.

Caravenna, F. (2005a). Random Walk Models and Probabilistic Techniques for Inhomogeneous Polymer Chains, *Ph.D. Thesis*, Univ. Paris 7 (F) and Univ. di Milano-Bicocca (I), *arXiv.org e-Print archive*: math.PR 0511561

Caravenna, F. (2005b). A Local Limit Theorem for Random Walks Conditioned to Stay Positive, *Probab. Theory Relat. Fields* **133**, pp. 508–530.

Caravenna, F. and Giacomin, G. (2005). On Constrained Annealed Bounds for Pinning and Wetting Models, *Elect. Comm. Probab.* **10**, pp. 179–189.

Caravenna, F., Giacomin, G. and Gubinelli, M. (2006). A Numerical Approach to Copolymers at Selective Interfaces, *J. Statist. Phys.* **122**, pp. 799–832.

Caravenna, F., Giacomin, G. and Zambotti, L. (2005). A Renewal Theory Approach to Copolymer Models with Adsorption, *arXiv.org e-Print archive*: math.PR 0507178

Caravenna, F., Giacomin, G. and Zambotti, L. (2006a). Sharp Asymptotic Behavior for Wetting Models in $(1 + 1)$-dimension, *Elect. J. Probab.* **11**, pp. 345–362.

Caravenna, F., Giacomin, G. and Zambotti, L. (2006b). Infinite Volume Limits of Polymer Chains with Periodic Charges, *arXiv.org e-Print archive*: math.PR 0604426

Causo, M. S. and Whittington, S. G. (2003). A Monte Carlo Investigation of the Localization Transition in Random Copolymers at an Interface, *J. Phys. A: Math. Gen.* **36**, pp. L189–L195.

Chayes, J. T., Chayes, L., Fisher D. S. and Spencer, T. (1986). Finite-Size Scaling and Correlation Lengths for Disordered Systems, *Phys. Rev. Lett.* **57**, pp. 2999–3002.

Chayes, J. T., Chayes, L., Fisher D. S. and Spencer, T. (1989). Correlation Length Bounds for Disordered Ising Ferromagnets, *Commun. Math. Phys.* **120**, pp. 501–523.

Cheliotis, D. (2005). Diffusion in Random Environment and the Renewal Theorem, *Ann. Probab.* **33**, pp. 1760–1781.

Coluzzi, B. (2005). Numerical Study on a Disordered Model for DNA Denaturation Transition, *Phys. Rev. E* **73**, 011911 (9 pages).

Comets, F., Shiga, T. and Yoshida, N. (2004). Probabilistic Analysis of Directed Polymers in Random Environment: a Review, *Advanced Studies in Pure Mathematics* **39**, pp. 115–142.

Cule, D. and Hwa, T. (1997). Denaturation of Heterogeneous DNA, *Phys. Rev. Lett.* **79**, pp. 2375–2378.

Dembo, A. and Zeitouni, O. (1998). Large Deviations Techniques and Applications, Second edition. Applications of Mathematics (New York), **38**. Springer-Verlag, New York.

Derrida, B., Hakim, V. and Vannimenius, J. (1992). Effect of Disorder on Two-dimensional Wetting, *J. Statist. Phys.* **66** (1992), pp. 1189–1213.

Deuschel, J.-D., Giacomin, G. and Zambotti, L. (2005). Scaling Limits of Equilibrium Wetting Models in (1+1)-dimension, *Probab. Theory Rel. Fields*, **132**, 471–500.

Deuschel, J.-D. and Stroock, D. W. (1989). Large Deviations, Pure and Applied Mathematics, **137**. Academic Press, Inc., Boston, MA.

Doney, R. A. (1997). One-sided Local Large Deviation and Renewal Theorems in the Case of Infinite Mean, Probab. Theory Relat. Fields **107**, pp. 451–465.

von Dreifus, H., Klein, A. and Fernando Perez, J. (1995). Taming Griffiths' Singularities: Infinite Differentiability of Quenched Correlation Functions, *Comm. Math. Phys.* **170**, pp. 21–39.

Feller, W. (1966). An Introduction to Probability Theory and its Applications, vol. 1, 2nd edition, Wiley series in probability and mathematical statistics, John Wiley & Sons. Inc., New York–London–Sydney.

Feller, W. (1971) An Introduction to Probability Theory and its Applications, Vol. II, Second edition, John Wiley & Sons, Inc., New York–London–Sydney.

Fisher, M. E. (1984). Walks, Walls, Wetting, and Melting, *J. Statist. Phys.* **34**, pp. 667–729.

Fisher, D. S. (1992). Random Transverse Field Ising Spin Chains, *Phys. Rev. Lett.* **69**, pp. 534–537.

Fitzsimmons, P. J., Fristedt, B. and Maisonneuve B. (1985). Intersections and Limits of Regenerative Sets, *Z. Wahr. verw. Geb.* **70**, pp. 157–173.

Fixman, M. and Freire, J. J. (1977). Theory of DNA Melting Curves, *Biopolymers* **16**, pp. 2693–2704.

Flory, P. J. (1953). Principles of Polymer Chemistry, Cornell University Press, Ithaca (NY).

Forgacs, G., Luck, J. M., Nieuwenhuizen, Th. M. and Orland, H. (1986). Wetting of a Disordered Substrate: Exact Critical Behavior in Two Dimensions, *Phys. Rev. Lett.* **57**, pp. 2184–2187.

Funaki, T. (2005). Concentration Property for Wiener Measure with Density Having Two Large Deviation Minimizers, *preprint*, available on the webpage of the author.

Galluccio, S. and and Graber, R. (1996). Depinning Transition of a Directed Polymer by a Periodic Potential: a d-dimensional Solution, *Phys. Rev. E* **53**, pp. R5584–R5587.

Garel, T., Huse, D. A., Leibler, S. and Orland, H. (1989). Localization Transition of Random Chains at Interfaces, *Europhys. Lett.* **8**, 9–13.

Garel, T. and Monthus, C. (2005a). Numerical Study of the Disordered Poland-Scheraga Model of DNA Denaturation, *J. Stat. Mech., Theory and Experiments*, P06004.

Garel, T. and Monthus, C. (2005b). Distribution of Pseudo-critical Temperatures and Lack of Self-averaging in Disordered Poland-Scheraga Models with Different Loop Exponents, *Eur. Phys. J. B* **48**, pp. 393–403.

Garel, T. and Monthus, C. (2005c). Delocalization Transition of the Selective Interface Model: Distribution of Pseudo-critical Temperatures, *J. Stat. Mech., Theory and Experiments*, P12011.

Garel, T., Monthus, C. and Orland, H. (2000). Copolymer at a Selective Interface and Two-dimensional Wetting: a Grand Canonical Approach, *Eur. Phys. J. B* **17**, pp. 121–130.

Garsia, A. and Lamperti, J. (1963). A Discrete Renewal Theorem with Infinite Mean, Comment. Math. Helv. **37**, pp. 221–234.

de Gennes, P. G. (1979). Scaling Concepts in Polymer Physics, Cornell University Press, Ithaca (NY).

Georgii, H.-O. (1988). Gibbs Measures and Phase Transitions, de Gruyter Studies in Mathematics **9**, Walter de Gruyter & Co., Berlin.

Giacomin, G. (2004). Localization Phenomena in Random Polymer Models, *unpublished notes*, available on the webpage of the author.

Giacomin, G. and Toninelli, F. L. (2005). Estimates on Path Delocalization for Copolymers at Selective Interfaces, *Probab. Theory Rel. Fields* **133**, pp. 464–482.

Giacomin, G. and Toninelli, F. L. (2006a). Smoothing Effect of Quenched Disorder on Polymer Depinning Transitions, *Commun. Math. Phys.* **266**, pp. 1–16.

Giacomin, G. and Toninelli, F. L. (2006b). The Localized Phase of Disordered Copolymers with Adsorption, *Alea* **1**, 149–180.

Giacomin, G. and Toninelli, F. L. (2006c). Smoothing of Depinning Transitions for Directed Polymers with Quenched Disorder, *Phys. Rev. Lett.* **96**, 070602.

Guerra, F. and Toninelli, F.L. (2002). The Thermodynamic Limit in Mean Field Spin Glass Models, *Commun. Math. Phys.* **230**, pp. 71–79.

Habibzadah, N., Iliev, G. K., Martin, R., Saguia A. and Whittington, S. G., (2006). The Order of the Localization Transition for a Random Copolymer,

J. Phys. A: Math. Gen. **39**, pp. 5659–5667.

Harris, A. B. (1974). Effect of Random Defects on the Critical Behaviour of Ising Models, *J. Phys. C* **7**, pp. 1671–1692.

den Hollander, F. and Whittington, S. G. (2006). Localization Transition for a Copolymer in an Emulsion, *arXiv.org e-Print archive*: math.PR 0512374

den Hollander, F. and Wüthrich, M. (2004). Diffusion of a Heteropolymer in a Multi-interface Medium. *J. Statist. Phys.* **114**, pp. 849–889.

Hryniv, O. and Velenik, Y. (2004). Universality of Critical Behaviour in a Class of Recurrent Random Walks, *Probab. Theory Relat. Fields* **130**, pp. 222–258.

Iliev, G., Orlandini, E. and Whittington, S. G. (2004). Adsorption and Localization of Random Copolymers Subject to a Force: The Morita Approximation, *Eur. Phys. J. B* **40**, pp. 63–71.

Iliev, G., Rechnitzer, A. and Whittington, S. G. (2005). Localization of Random Copolymers and the Morita Approximation, *J. Phys. A: Math. Gen.* **38**, pp. 1209–1223.

Imry, Y. and Ma, S.-K. (1975). Random-Field Instability of the Ordered State of Continuous Symmetry, *Phys. Rev. Lett.* **35**, pp. 1399–1401.

Isozaki, Y. and Yoshida, N. (2001). Weakly Pinned Random Walk on the Wall: Pathwise Descriptions of the Phase Transition, *Stoch. Proc. Appl.* **96**, pp. 261–284.

Jain, N. C. and Pruitt, W. E. (1972). The Range of Random Walk, in *Proceedings of the Sixth Berkeley Symposium on Mathematical Statistics and Probability* (Univ. California, Berkeley, Calif., 1970/1971), Vol. III: Probability theory, pp. 31–50. Univ. California Press, Berkeley, Calif., 1972.

Janse van Rensburg, E. J. , Rechnitzer, A., Causo, M. S., and Whittington, S.G. (2001). Self-averaging Sequences in the Statistical Mechanics of Random Copolymers, *J. Phys. A: Math. Gen.* **34**, pp. 6381–6388.

Janvresse, E., de la Rue, T. and Velenik, Y. (2005). Pinning by a Sparse Potential, *Stoch. Proc. Appl.*, **115**, 1323–1331.

Jenkins, A. D., Kratochvíl, P., Stepto, R. F. T. and Suter, U. W. (1996). Glossary of Basic Terms in Polymer Sciences, *Pure & Appl. Chem.* **68**, pp. 2287–2311.

Kafri, Y., Mukamel, D. and Peliti, L. (2000). Why is the DNA Denaturation Transition First Order?, *Phys. Rev. Lett.* **85**, pp. 4988–4991.

Kafri, Y., Nelson, D. R. and Polkovnikov, A. (2006). Unzipping Flux Lines from Extended Defects in Type-II Superconductors, Europhys. Lett. **73**, pp. 253–259.

Kesten, H. (1963). Ratio Theorems for Random Walks II, *J. Analyse Math.* **11**, pp. 323-379.

Kingman, J. F. C. (1961). A Convexity Property of Positive Matrices, *Quart. J. Math.* **12**, pp. 283–284.

Kingman, J. F. C. (1973). Subadditive Ergodic Theory, *Ann. Probab.* **1**, pp. 882–909.

Krengel, U. (1985). Ergodic Theorems, de Gruyter Studies in Mathematics **6**, Walter de Gruyter & Co..

Kühn, R. (1996). Equilibrium Ensemble Approach to Disordered Systems I: General Theory, Exact Results, *Z. Phys. B* **100**, pp. 231–242.

Le Doussal, P., Monthus, C. and Fisher, D. S. (1999). Random Walkers in One-dimensional Random Environments: Exact Renormalization Group Analysis, *Phys. Rev. E* **59** (1999), pp. 4795–4840.

Ledoux, M. (2001). The Concentration of Measure Phenomenon, *Mathematical Surveys and Monographs*, **89**, American Mathematical Society, Providence, RI.

van Leeuwen, J. M. J. and Hilhorst, H. J. (1981). Pinning of Rough Interface by an External Potential, *Phys. A* **107**, pp. 319–329.

Liggett, T. M. (1985). An Improved Subadditive Ergodic Theorem, *Ann. Probab.* **13**, pp. 1279–1285.

Lubensky, D. K. and Nelson, D. R. (2000). Pulling Pinned Polymers and Unzipping DNA, *Phys. Rev. Lett.* **85**, pp. 1572–1575.

Lund, R. B. and Tweedie, R. L. (1996). Geometric Convergence Rates of Stochastically Ordered Markov Chains, *Math. Oper. Research* **21**, pp. 182–194.

Madras, N. (2002). Lectures on Monte Carlo Methods, *Fields Institute Monographs* **16**, American Mathematical Society, Providence (RI).

Madras, N. and Slade, G. (1993). The Self-Avoiding Walk, Probability and its Applications, Birkhäuser, Boston (MA).

Madras, N. and Whittington, S. G. (2003). Localization of a Random Copolymer at an Interface, *J. Phys. A: Math. Gen.* **36**, 923–938.

Marenduzzo, D., Trovato, A. and Maritan, A. (2001). Phase Diagram of Force-induced DNA Unzipping in Exactly Solvable Models, *Phys. Rev. E* **64**, 031901 (12 pages).

Maritan, A., Riva, M. P. and Trovato, A. (1999). Heteropolymers in a Solvent at an Interface, *J. Phys. A: Math. Gen.* **32**, pp. L275–L280.

Matheron, G. (1975). Random Sets and Integral Geometry, John Wiley & Sons., New York, NY.

Matsumoto, M. and Nishimura, T. (1998). Mersenne Twister: A 623-dimensionally Equidistributed Uniform Pseudorandom Number Generator, *ACM Trans. on Mod. and Comp. Simul.* **8**, pp. 3–30.

Minc, H. (1988). Nonnegative Matrices, *Wiley-Interscience Series in Discrete Mathematics and Optimization*, John Wiley & Sons, Inc., New York.

Monthus, C. (2000). On the Localization of Random Heteropolymers at the Interface Between Two Selective Solvents, *Eur. Phys. J. B* **13**, pp. 111–130.

Morita, T. (1966). Statistical Mechanics of Quenched Solid Solutions with Application to Magnetically Dilute Alloys, *J. Math. Phys.* **5**, pp. 1401–1405.

Naidedov, A. and Nechaev, S. (2001). Adsorption of a Random Heteropolymer at a Potential Well Revisited: Location of Transition Point and Design of Sequences, *J. Phys. A: Math. Gen.* **34**, pp. 5625–5634.

Nechaev, S. and Zhang, Y.–C. (1995). Exact Solution of the 2D Wetting Problem in a Periodic Potential, *Phys. Rev. Lett.* **74**, pp. 1815–1818.

Ney, P. (1981). A Refinement of the Coupling Method in Renewal Theory, *Stoch. Proc. Appl.* **11**, pp. 11–26.

Olivieri, E. and Vares, M. E. (2005). Large Deviations and Metastability, *Encyclopedia of Mathematics and its Applications* **100**, Cambridge University Press, Cambridge (UK).

Orlandini, E., Tesi, M. C. and Whittington, S. G. (2000). Self-averaging in Models of Random Copolymer Collapse, *J. Phys. A* **33**, 259–266.

Orlandini, E., Tesi, M. C. and Whittington, S. G. (2002). Self-averaging in the Statistical Mechanics of Some Lattice Models, *J. Phys. A* **35**, pp. 4219–4227.

Orlandini, E., Tesi, M. C. and Whittington, S. G. (2004). Adsorption of a Directed Polymer Subject to an Elongational Force, *J. Phys. A: Math. Gen.* **37**, pp. 1535–1543.

Peng, C. K., Buldyrev, S. V., Goldberger, A.L., Havlin, S., Sciortino, F., Simons, M. and Stanley, H. E. (1992). Long-Range Correlations in Nucleotide Sequences, *Nature* 1992 **356**, pp. 168–170.

Pétrélis, N. (2005). Polymer Pinning at an Interface, *arXiv.org e–Print archive*: math.PR 0504464, *Stoch. Proc. Appl.* (to appear).

Pétrélis, N. (2006). Localisation d'un Polymère en Interaction avec une Interface, *Ph.D. Thesis*, Univ. de Rouen (F).

Petrov, V. V. (1975). Sums of Independent Random Variables, *Ergebnisse der Mathematik und ihrer Grenzgebiete* **82**, Springer-Verlag, New York–Heidelberg.

Poland, D. (1974). Recursion Relation Generation of Probability Profiles for Specific-Sequence Macromolecules with Long-Range Correlations, *Biopolymers* **13**, pp. 1859–1871.

Port, S. C. (1963). An Elementary Probability Approach to Fluctuation Theory, *J. Math. Anal. Appl.* **6**, pp. 109–151.

R Development Core Team (2004). R: A Language and Environment for Statistical Computing, R Foundation for Statistical Computing, Vienna, Austria, ISBN 3-900051-07-0. URL: http://www.R-project.org

Rechnitzer, A. and Janse van Rensburg, E. J. (2004). Exchange Relations, Dyck Paths and Copolymer Adsorption, *Discrete Appl. Math.* **140**, pp. 49–71.

Revuz, D. and Yor, M. (1999). Continuous martingales and Brownian motion. Third edition. *Grundlehren der Mathematischen Wissenschaften* **293**, Springer-Verlag, Berlin.

Richard, C. and Guttmann, A. J. (2004). Poland-Scheraga Models and the DNA Denaturation Transition, *J. Statist. Phys.* **115**, pp. 943–965.

Roynette, B., Vallois, P. and Yor, M. (2003). Limiting Laws Associated with Brownian Motion Perturbated by Normalized Exponential Weights, *C. R. Math. Acad. Sci. Paris* **337**, pp. 667–673.

Schäfer, L. (2005). Can Finite Size Effects in the Poland-Scheraga Model Explain Simulations of a Simple Model for DNA Denaturation? *arXiv.org e-Print archive*: cond-mat 0502668

Sinai, Ya. G. (1993). A Random Walk with a Random Potential, *Theory Probab. Appl.* **38**, pp. 382–385.

Sinai, Ya. G. and Spohn, H. (1996). Remarks on the Delocalization Transition for Heteropolymers, in *Topics in Statistical and Theoretical Physics*, F. A. Berezin Memorial Volume, R. L. Dobrushin et al. Eds., AMS Transl. **177**, pp. 219–223.

Sommer, J.-U. and Daoud, M. (1995). Copolymers at Selective Interfaces, *Euro-*

phys. Lett., **32**, pp. 407–412.

Sommer, J.-U. and Daoud, M. (1996). Adsorption of Multiblock Copolymers at Interfaces between Selective Solvents: Single-chain Properties, *Phys. Rev. E* **53**, pp. 905–920.

Soteros, C. E. and Whittington, S. G. (2004). The Statistical Mechanics of Random Copolymers, *J. Phys. A: Math. Gen.* **37**, R279–R325.

Stepanow, S., Sommer, J.-U. and Erukhimovich, I. Ya. (1998). Localization Transition of Random Copolymers at Interfaces, *Phys. Rev. Lett.* **81**, pp. 4412–4416.

Sznitman, A.-S. (1998). Brownian Motion, Obstacles and Random Media; Springer Monographs in Mathematics. Springer-Verlag, Berlin (D).

Talagrand, M. (1996). A New Look at Independence, *Ann. Probab.* **24**, 1–34.

Talagrand, M. (2003). Spin Glasses: a Challenge for Mathematicians. Cavity and Mean Field Models, *A Series of Modern Surveys in Mathematics*, Springer-Verlag, Berlin (D).

Tang, L.-H. and Chaté, H. (2001). Rare-Event Induced Binding Transition of Heteropolymers, *Phys. Rev. Lett.* **86**, pp. 830–833.

Toninelli, F. L. (2006). Critical Properties and Finite Size Estimates for the Depinning Transition of Directed Random Polymers, *J. Statist. Phys.* (to appear).

Trovato, A. and A. Maritan, A. (1999). A Variational Approach to the Localization Transition of Heteropolymers at Interfaces, *Europhys. Lett.* **46**, pp. 301–306.

Upton, P. J. (1999). Exact Interface Model for Wetting in the Planar Ising Model, *Phys. Rev. E* **60**, pp. R3475–R3478.

Velenik, Y. (2006). Localization and Delocalization of Random Interfaces, *Probability Surveys* **3**, pp. 112–169.

Villani, C. (2003). Topics in Optimal Transportation, *Graduate Studies in Mathematics*, **58**, American Mathematical Society, Providence, RI.

Whittington, S. G. (1998). A Directed-walk Model of Copolymer Adsorption, *J. Phys. A: Math. Gen.* **31**, pp. 8797–8803.

Williams, D. (1991). Probability with Martingales, *Cambridge Mathematical Textbooks*, Cambridge University Press (UK).

Zeitouni, O. (2004). Random Walks in Random Environment, Lectures on probability theory and statistics (Saint-Flour 2001), pp. 193–312, *Lecture Notes in Mathematics* **1837**, Springer, Berlin (D).

Index